Resilience Reset

Drawing on evidence from urban resilience initiatives around the globe, the authors make a compelling argument for a "resilience reset", a pause and stocktake that critically examines the concepts, practices and challenges of building resilience, particularly in cities of the Global South.

In turn, the book calls for the world's cities to alter their course and "pivot" towards novel approaches to enhancing resilience. The book presents shifts in ways of acquiring and analysing data, building community resilience, approaching urban planning, engaging with informality, delivering financing and building the skills of those running cities in a post-COVID world grappling with climate-related impacts. In *Resilience Reset*, the authors encourage researchers, policymakers and practitioners to break out of existing modes of thinking and doing that may no longer be relevant for our rapidly urbanising and dynamic world. The book draws on the latest academic and practice-based evidence to provide actionable insights for cities that will enable them to deal with multiple interacting shocks and stresses.

The book will be an indispensable resource to those studying urbanisation, development, climate change and risk management as well as for those designing and deploying operational initiatives to enhance urban resilience in businesses, international organisations, civil society organisations and governments. It is a must-read for anyone interested in managing the risks of climate impacts in urban centres in the Global South.

Aditya V. Bahadur, PhD, is a Principal Researcher at the International Institute for Environment and Development specialising in climate-resilient urban development. Previously he worked with the Action on Climate Today programme supporting governments with climate policy and institutional development and with the ODI (Overseas Development Institute). Aditya completed a PhD at the University of Sussex, UK, and a post-doctoral programme at Columbia University, USA.

Thomas Tanner, PhD, is a development geographer specialising in building resilience and adaptation to climate change through development policy and practice. He is Director of the Centre for Development, Environment and Policy in the Department of Development Studies at SOAS University of London, where he leads work on climate change, sustainability and development. He previously held positions at the ODI, the Institute of Development Studies, the United Nations Development Programme and the UK's Department for International Development.

Resilience Reset

Creating Resilient Cities in the Global South

Aditya V. Bahadur
and Thomas Tanner

Routledge
Taylor & Francis Group
LONDON AND NEW YORK

First published 2022
by Routledge
2 Park Square, Milton Park, Abingdon, Oxon OX14 4RN

and by Routledge
605 Third Avenue, New York, NY 10158

Routledge is an imprint of the Taylor & Francis Group, an informa business

British Library Cataloguing-in-Publication Data
A catalogue record for this book is available from the British Library

Library of Congress Cataloging-in-Publication Data
Names: Bahadur, Aditya V., author. | Tanner, Thomas, 1975– author.
Title: Resilience reset : creating resilient cities in the Global South /
Aditya V. Bahadur and Thomas Tanner.
Description: Abingdon, Oxon ; New York, NY : Routledge, 2021. |
Includes bibliographical references and index.
Identifiers: LCCN 2020057325 (print) | LCCN 2020057326 (ebook) |
ISBN 9780367375485 (hardback) | ISBN 9780367375508 (paperback) |
ISBN 9780429355066 (ebook)
Subjects: LCSH: Resilience (Personality trait)–Social aspects. | Urbanization.
Classification: LCC BF698.35.R47 B36 2021 (print) |
LCC BF698.35.R47 (ebook) | DDC 155.2–dc23
LC record available at https://lccn.loc.gov/2020057325
LC ebook record available at https://lccn.loc.gov/2020057326

ISBN: 978-0-367-37548-5 (hbk)
ISBN: 978-0-367-37550-8 (pbk)
ISBN: 978-0-429-35506-6 (ebk)

Typeset in Times New Roman
by Newgen Publishing UK

Contents

Figures

Boxes

Tables

Foreword

OVER a space of two decades, the concept of urban resilience has evolved from being a specialist concern of engineers and disaster managers to becoming a paradigm for urban governance across a broad range of sectors and systems. This progression has been supported by substantial investments stimulated by the growing risks presented by human-induced climate change. Focusing particularly on the Global South, urban climate change resilience (UCCR) has received significant attention from international researchers and policymakers alike. This concept is now a mature formulation that finds wide resonance among the urban development, risk management and international development communities of practice.

The time is therefore ripe to pause, take stock and reflect critically on the effectiveness of the dominant modes, mechanisms and entry points for building UCCR that have been employed across the world. This is why I was delighted when the authors of this book introduced me to their key arguments. This is a timely volume that maps the existing state of play regarding UCCR, examines the major challenges, and proposes new directions for overcoming them. It is a must-read for anyone interested in managing the risks of climate impacts in urban centres, not only in the Global South but throughout the world.

The arguments in this book are highly topical because more people are living in cities than ever before and because the impacts of climate change are intensifying. Be it catastrophic floods in Chennai, water scarcity in Cape Town, cyclones in New Orleans or heatwaves in Paris, the last two decades have demonstrated that the battle against climate risk will be lost or won in the world's cities. This book tackles this challenge head-on and encourages researchers, policymakers and practitioners to break out of existing modes of thinking and action that may no longer be relevant for our rapidly urbanising and dynamic world.

As well as summing up the state of play in key areas, the authors present a critical analysis and forward-looking arguments. The book tackles a series of impasses in the contexts of information and data, community resilience, planning, urban service delivery and financing. In each area, the book sets forth crucial directions to which thinking and action must "pivot" if they are to drive equitable and effective resilience in cities. Respectively, their calls for the exploration of big data, transformative approaches, informality, soft systems and endogenous financing will find audiences globally. As such, the book provides a comprehensive foundation in relevant, contemporary debates on UCCR.

Taken together, this exploration of the frontiers of the practice, research and policy of urban resilience is a timely reminder that reflection and renewal are essential for sustaining the well-being of urban residents in the face of climate change. It is inspiring that this analysis goes beyond esoteric examination of meanings and definitions to provide actionable insights for cities as they move forward in developing their climate change policies and programmes.

The members of the Urban Climate Change Research Network (UCCRN), as well as all readers, are sure to find this book rewarding. I have no doubt that it will be recognised as an important contribution to the field of knowledge on urban climate change resilience.

Dr Cynthia Rosenzweig
Co-Director, Urban Climate Change Research Network
New York, June 2020

Preface

THIS book draws largely on evidence around urban resilience from before the onset of the COVID-19 pandemic. The authors have taken stock and reviewed insights in light of this dramatic global change. We contend that the crisis to a large extent reinforces the challenges and pivots we identify. While we do not attempt to provide a full analysis of the pandemic and its influence, we have enhanced sections of the key chapters with insights into emerging synergies: COVID-19 is pushing city governments to embrace the use of big data for reducing risk and to expand partnerships with the informal sector, and novel forms of endogenously driven financing are more crucial than ever. The final chapter distils in more detail the implications of the book's main findings for managing multiple interacting shocks and stresses in the future. We optimistically see the potential to tackle the post-COVID recovery through these pivots for urban resilience. From disruption comes the potential for change, and we hope this book makes a contribution to future cities that can adapt and thrive to meet future challenges and opportunities.

Acknowledgements

THIS book would never have been possible without the unwavering support of our families, friends and colleagues. We are deeply grateful to our wonderful partners Aditi and Gemma, who saw us through many book-filled nights and weekends. Thanks also to Minakshi and Shashivansh Bahadur and Ruth and Brian Tanner for their valuable help with parts in the process.

This book would not have been possible without the generosity of the Fulbright-Kalam Climate Fellowship for Postdoctoral Research administered by the United States-India Educational Foundation (Fulbright Commission). It is also the result of the incredible support of Cynthia Rosenzweig at the NASA Goddard Institute for Space Studies, New York, who helped shape the key arguments in the book through her commanding insight into climate change and cities. The book also benefited from invaluable advice and help from Somayya Ali Ibrahim and the Urban Climate Change Research Network.

We benefited greatly from the input and advice of a range of people. Thanks in particular to Cristina Rumbaitis del Rio, Tom Mitchell, David Dodman, Saleemul Huq, Marcus Moench, Clare Shakya, Ammar Malik, John Firth, Sarah Opitz-Stapleton, Daniel Zarrilli, Anu Jogesh, Lindsey Jones, Eric Kasper and Simon Addison.

We would also like to acknowledge others those whose work has inspired and informed us in the process of writing the book, including Adriana Allen, Diane Archer, Amir Bazaz, Mihir Bhatt, Christian V. Braneon, Nick Brooks, Anna Brown, Harriet Bulkeley, Terry Cannon, Eric Chu, Patrick Cobbinah, Jo da Silva, Simon Croxton, Stephen Devereux, Ashvin

Dayal, Carle Folke, Richard Friend, Elizabeth Gogoi, C. S. Holling, Maggie Ibrahim, Garima Jain, Cassidy Johnson, Sam Kernaghan, Fawad Khan, Kamal Kishore, Bijay Kumar, Shuaib Lwasa, Melissa Leach, Kenneth MacClune, Bernard Manyena, Heather McGray, Sara Meerow, Diana Mitlin, Karen O'Brien, Mark Pelling, Katie Peters, Umamaheshwaran Rajasekar, Aromar Revi, Ananya Roy, David Satterthwaite, Lisa Schipper, Ian Scoones, Rajib Shaw, Jo da Silva, Chandni Singh, Andrew Sumner, Anna Taylor, Maarten van Aalst, Shiraz Wajih, Emily Wilkinson and Gina Ziervogel.

Chapter 1

Urban climate change resilience "reset"

FOR the first time in the history of the world, more people live in towns and cities than in rural areas. Between 1950 and 2018, the share of the world's population living in urban areas almost doubled. By 2030, 6 out of every 10 people on Earth will be city dwellers. The most urbanised geographic regions include Northern America (82% urban), Latin America and the Caribbean (81%), Europe (74%) and Oceania (68%) (UN DESA 2018). Asia's population is 50% urban and Africa's is 43%. Some of the world's least economically developed regions have the fastest rates of urbanisation in the world. Low-income countries are growing at 4% every year, middle-income countries at 2.1% and high-income countries at 0.7% (World Bank 2020). About half of the world's urban residents live in cities of less than 500,000 people, while one in every eight lives in 33 "megacities" of over 10 million people. By 2030, there will be 43 megacities in the world, with an overwhelming majority of these in the Global South. Urbanisation is the defining demographic trend of our times.

A disproportionate number of the world's urban centres are located along coasts and rivers – locations that are highly exposed to climate change and disaster hazards. Two of every three urban residents in the world currently lives along a coast and this will rise to three of four urban residents by 2025 (Gencer et al. 2018). Under higher emissions greenhouse gas scenarios and with early-onset Antarctic ice sheet instability, 21st-century sea level rise may exceed 2 metres (Wong et al. 2017). Even with sharp cuts to greenhouse gas emissions, 570 low-lying coastal cities could see the sea level rise by an average of half a metre, leading to 800 million being put at risk and more than a trillion dollars' worth of damage to property (Rosenzweig and Solecki 2018). Alongside coastal storms and sea level rise, "a number of large urban agglomerations in almost all continents will be exposed to a temperature rise of greater than 1.5°C" and greater heat extremes as a result (Revi et al. 2014: 554). In the three decades between 2000 and 2030 the amount of urban area exposed to flood and drought will increase by 250% (Güneralp et al. 2015). These hazards pose a major challenge. Some 1 billion urban residents live in contexts characterised by a high degree of vulnerability and low adaptive capacity, where there is inadequate access to healthcare, emergency services, policing/rule of law, protective infrastructure, and healthy and safe housing and where the occurrence of disasters from extreme weather is very common (Revi et al. 2014). The shocks and stresses from a changing climate are a major challenge for urban areas across the world.

The high vulnerability and increasing exposure of urban populations combines with growing climate-related hazards to present a daunting climate risk profile for many of the world's cities. However, as we have gained a better understanding of these dynamics and experienced more climate-related shocks, there has been a growing global interest in policy responses to enhance urban resilience and adapt to future change. The urgency of tackling the global COVID-19 pandemic has added further to the impetus for managing climate risk in the context of wider risks to urban areas. This chapter provides a framing of urban resilience contexts and evolution in thinking and practice. While the book focuses its attentions on cities of the Global South, we draw on wider geographical contexts, arguing that significant lessons can be learned but also that contexts render translation of blueprints problematic. Setting up the other chapters in the book, we argue that the current patterns of development in urban centres and evolving risk context is rendering existing systems of risk management redundant. The book therefore outlines a set of "pivots" towards novel approaches for enhancing resilience in five specific areas of action.

1.1 THE RESILIENCE CONTEXT

The use of "resilience thinking" as a conceptual paradigm has risen sharply across multiple disciplines in books, scholarly journals and scientific research. A Web of Science search shows that there was a 25-fold increase in the use of the term "resilience" in journal articles between 1998 and 2018. Interest in the word "resilience" on Google has almost quadrupled between 2010 and 2020 (Google Trends 2020). Even though the term is now dominant within the fields of climate change and disasters, it is used by researchers and practitioners in a range of fields spanning psychology, economics, engineering, organisational change and ecology.

Resilience as it is explored in ecology and, more specifically, the study of social-ecological systems, has been influential in informing responses to climate change impacts in policy and practice. Here, resilience is defined as the "measure of the persistence of systems and of their ability to absorb change and disturbance and still maintain the same relationships between populations or state variables" (Holling 1973: 14). This concept was originally explored in the context of ecosystems management drawing on, for instance, the study of lake systems and protected forests. These were studied as systems with a number of interconnected parts arranged in dynamic configurations, with resilience as the extent to which these systems could absorb shocks and stresses and still remain in approximately the same configuration. For instance, limnologists studied pollution flow into freshwater lakes to determine the amount that a lake could take before it transformed into a different state (e.g. into a eutrophied or murky state) (Holling 1973). Over time this evolved into the study of coupled social-ecological systems, or "linked systems of people and nature. The term emphasises that humans must be seen as a part of, not apart from, nature – that the delineation between social and ecological systems is artificial and arbitrary" (Simonsen 2007).

This understanding of resilience has been translated into a number of principles or tenets that can be applied in the management of social-ecological systems. These include maintaining diversity and redundancy (where systems have multiple components and built-in buffer capacity); managing connectivity (where systems should be able to network or decouple from other systems); managing slow variables and feedbacks (where the relationship between variables in a system can be reinforced or dampened); fostering complex adaptive systems thinking (where management approaches need to shift with changing circumstances); encouraging learning (existing knowledge needs to be revised continually); broadening participation (where multiple perspectives are included in decision-making); and promoting polycentric governance systems (where multiple governing bodies interact to make and enforce rules) (Simonsen et al. 2015).

1.1.1 Urban resilience

Urban resilience broadly refers to the ability of cities and towns to function and adapt in the face of shocks and stresses (Leichenko 2011; Wardekker et al. 2010; Silva et al. 2012; Satterthwaite and Dodman 2013). We employ the definition from a systematic review where urban resilience refers to

> *the ability of an urban system-and all its constituent socio-ecological and socio-technical networks across temporal and spatial scales-to maintain or rapidly return to desired functions in the face of a disturbance, to adapt to change, and to quickly transform systems that limit current or future adaptive capacity.*
>
> (Meerow et al. 2016: 39, italics in original)

Urban climate change resilience is more focused and concerns itself with disturbances that are primarily hydrometeorological. While many initiatives have been prompted by the threats of climate change, many responses have encompassed resilience to a wider range of risks. This is particularly salient in the context of the opportunities for transformational responses to the COVID-19 pandemic that can generate more climate-resilient development pathways (Schipper et al. 2020).

Thinking on urban resilience has evolved since it rose to prominence at the end of the 20th century . We distinguish three phases of this evolution based on an analysis of literature (see Table 1.1; Bahadur and Thornton 2015).

In the first phase, urban resilience refers largely to terrorism, security concerns or shocks; rarely does it engage with long-term stresses and change or make links to climate change-related risks. Events such as the 9/11 terrorist attacks in New York in 2001 and Hurricane Katrina in New Orleans in 2005 led to a spate of research on cities, security risks and natural hazards (Godschalk 2003; Bull-Kamanga et al. 2003). Resilience is considered as a response to particular hazards rather than as an approach to managing functioning systems more broadly (Meerow et al. 2016). Much of the literature in this early phase is based on commentaries and secondary research rather than primary empirical

TABLE 1.1 The evolution of urban resilience

	Data	Disturbances	Conceptual Rigour
Phase 1	Lack of primary empirical data	Resilience seen as a specific response to shocks in urban areas	Resilience used without much reference to conceptual underpinnings
Phase 2	Turn towards the use of primary empirical data	Resilience starts to be seen as a management strategy to deal with multiple shocks and stresses in urban areas	Resilience used with references to underpin conceptual paradigms and heuristics
Phase 3	Substantial and extensive use of primary empirical data, including from urban resilience programmes	Resilience firmly acknowledged as a mode of urban governance necessary for dealing with a range of shocks and stresses	Considerable conceptual development, consolidation of urban climate change resilience as a paradigm and emergence of empirically informed analytical frameworks

data (Swanstrom 2008). There is limited engagement with the conceptual underpinnings of resilience, either in the context of social-ecological systems or other fields. As such, it is used largely without exploring analytical constructs or heuristics such as non-equilibrium dynamics, complex systems thinking and feedbacks (Bahadur and Thornton 2015).

In the second phase, urban resilience starts to plug many of the gaps from the first phase. Instead of looking at resilience as a specific response to particular shocks, papers start to consider resilience as management approach for dealing with a multitude of shocks and stresses (Evans 2011; Wardekker et al. 2010; Leichenko 2011; Gasper et al. 2011). Crucially, urban resilience starts to actively consider shocks and stresses emanating from climate change (Evans 2011). This phase witnesses a turn towards primary empirical data in exploring approaches for building resilience; there is extensive use of case study research and salient pieces of literature are located in diverse geographical contexts (Ernstson et al. 2010; Evans 2011; Wardekker et al. 2010; Pelling and Manuel-Navarrete 2011; Pelling 2011). The literature also starts to become incrementally more "practice oriented", and we see organisations developing tools, techniques and operational frameworks for enhancing resilience in urban areas (Brown and Kernaghan 2011). This reflects the launch at the end of the 2000s of major international development initiatives for resilience building (e.g. Rockefeller Foundation 2009). Finally, this phase begins to draw on the richer conceptual underpinnings of resilience thinking with urban exploration of complex systems, feedbacks between system components, thresholds and the adaptive cycle (Ernston et al. 2010; Pelling and Manuel-Navarrete 2011).

In the third phase, literature on urban resilience advances several trends, distinguishing urban climate change resilience as a subset of broader urban resilience (Tyler and Moench 2012). The focus now acknowledges building resilience to both shocks and stresses. Crucially, resilience is now also recognised as an urban governance approach, not just in terms of management practices to mitigate risk (Satterthwaite 2013; Friend and Moench 2013). Geographically situated studies and rich empirical data continue to proliferate, and there is a marked increase in the focus on towns and cities of developing countries (Prashar et al. 2012; Carmin et al. 2012; Button et al. 2013). A significant proportion of this literature draws on the work of the Asian Cities Climate Change Resilience Network (ACCCRN) – an international development initiative to enhance the resilience of urban institutions, systems and structures to climate risks and, through this, to measurably improve the lives of poor and vulnerable people (Rockefeller Foundation 2009; Friend and Moench 2013; Silva et al. 2012; Brown et al. 2012). This phase also sees new conceptual frameworks on urban climate change resilience. Tyler and Moench (2012: 319) draw on empirical data and concepts of resilience in social-ecological systems to present a framework, arguing that "resilience to climate change increases by building agent capacities, strengthening systems and reinforcing institutions that link agents and systems" (see Figure 1.1). Similarly, Rockefeller Foundation and Arup (2015) draws on empirical data to present a view of what a resilient city looks like and presents an outcome-oriented framework with 12 outcomes across four different categories and seven qualities of resilience (integrated, inclusive, resourceful, flexible, redundant, robust and reflective).

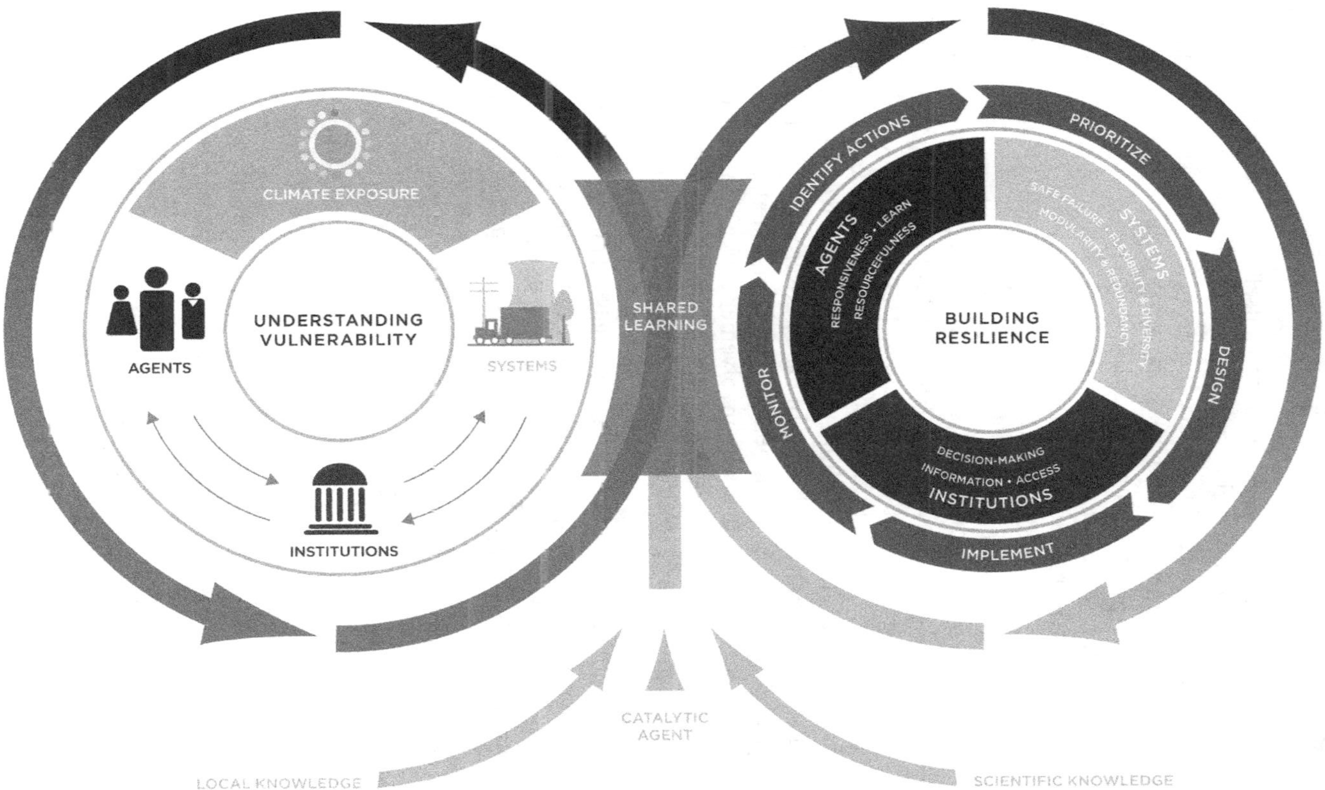

Figure 1.1 Framework for urban climate resilience. Source: Institute for Social and Environmental Transition-International (https://www.i-s-e-t.org)/Tyler and Moench (2012).

1.1.2 Urban resilience, systems thinking and complexity

As the concept of resilience concerns itself with the functioning of systems, any discussion of resilience must include an exploration of systems thinking and complexity. An analysis of salient literature in this domain reveals three main clusters of overlapping issues (see Bahadur et al. 2016).

First, a resilience lens acknowledges that urban areas are complex systems. Just as complex systems are an agglomeration of "interconnected and interdependent elements" (Ramalingam and Jones 2008: 8), urban areas are "sets of elements or components tied together through sets of interactions" (Batty 2008: 5). Any attempt to manage or govern cities can therefore draw on systems thinking and complexity, understanding that different economic, governance, social and infrastructural systems influence and interact with each other through constant feedbacks. In urban governance practice, this is evident in the establishment of platforms where different city actors converge and in the involvement of a wide variety of stakeholders, including the urban poor and marginalised, in decision-making (see Box 1.1) (Bahadur et al. 2016; Dodman and Carmin 2011).

Another factor that comes into focus when employing resilience to understand dynamics in urban areas is that changes at the level of the town or city are dependent on higher scales of governance. Social-ecological systems theorists have demonstrated that change at one spatial scale (e.g. farm) is influenced by changes at higher scales (e.g. regional watersheds); in the same way, the resilience of a town or city is influenced by subnational, national and international factors (Gunderson and Holling 2001). Feedbacks and dynamic interactions within economic, governance, social and infrastructural systems at multiple scales therefore need to be understood (Bahadur et al. 2016). For instance, the rate, scale and nature of physical urban development directly influences the degree to which people are exposed to climate risks, but this is highly dependent on national economic and planning regimes. To ascertain feedbacks across scales and ensure resilience, urban areas must be seen as complex systems that are nested within larger and broader complex systems.

Third, adopting systems thinking leads to an understanding of how towns and cities are part of a geographical continuum that stretches beyond traditional political boundaries (Bahadur et al. 2016). Enhancing urban resilience must therefore consider

BOX 1.1 City advisory committees as an instrument to manage urban complex systems

To promote understanding of cities as complex systems, the ACCCRN enabled the formation of city advisory committees. These provided a platform in which constituencies with a stake in the city's functioning were represented and charged with jointly understanding its vulnerabilities, developing and executing a resilience strategy and providing relevant information from all parts of the urban system. For instance, the committees formed in the cities of Indore and Gorakhpur in India saw active participation from the urban local bodies, the development authorities, local business people, groups representing marginalised urban residents, government officials charged with delivering urban services, civil society organisations, academics and meteorologists. These forums were used to undertake exercises such as "causal loop diagramming", which draws on the knowledge of diverse stakeholders to determine the links between seemingly disconnected issues (e.g. solid waste management and flooding). This permitted the development of a systemic understanding of the city and offered the opportunity to explore how its component parts would interact during the process of building resilience to climate change impacts.

Source: Bahadur (2014)

the influence of peri-urban areas, ex-urban areas, metropolitan regions and contiguous hinterland regions (Buxton et al. 2016). These geographies give and receive constant feedbacks from towns and cities. For example, peri-urban and hinterland areas provide vital ecosystem services, and metropolitan regions are essential for the smooth flow of goods, services and people to and from cities (Hedblom et al. 2017; Mani 2015). Urban systems interact with systems at higher scales and are influenced by areas lying beyond their immediate geographical limits.

1.2 THE URBAN CONTEXT

The key arguments in this book are anchored in a firm understanding of the urban context as well as the concept of resilience. Urban areas differ from non-urban areas due to their density, diversity, dynamism and complexity (Beall et al. 2010). These differences inform the conceptualisation and practice of urban resilience. Population density describes the number of people living in a specified land area, with urban density outstripping density in rural and hinterland areas. In some countries, the classification of particular areas as "urban" is based on population density. For instance, in India, urban areas must have a minimum population of 5,000, 75% of all male residents (of working age) should be engaged in non-agricultural livelihood activities, and there has to be a density of 400 people per square kilometre (Government of India 2011). Density of population, infrastructure and assets have major implications for risk and resilience. On the one hand, high density can increase risk, such as where proximity can lead to disease outbreaks. On the other hand, high density can aid resilience by, for example, enabling social networks or improving access for disaster response.

Urban contexts are generally more diverse than those of rural or hinterland areas. In the former, people from different classes, ethnic backgrounds, linguistic groups and professions reside in close proximity to one another. Delhi, India, contains close to 50 castes and at least four major linguistic groups (Government of India 2011); Johannesburg has been called "the world in one city"; and Jakarta is hailed as one of the most multicultural cities in the world (Hoon 2006). This diversity leads to people with very different social values and ways of viewing the world living close together, providing a sharp contrast to villages that have a greater degree of homogeneity. Diversity has implications for risk and resilience as it is known to fuel innovation (Beall et al. 2010), but it also poses potential barriers for the success of behaviour change and participatory decision-making processes essential for building resilience (Korf 2002; Bahadur and Tanner 2014).

Dynamism is another key characteristic of urban contexts. Towns and cities across the world, but especially in the Global South, are growing rapidly. In South Asia in the first decade of this century, the urban population grew by almost 130 million, equivalent to the combined populations of France and the United Kingdom (Ellis and Roberts 2016). In countries such as India, urbanisation is an overwhelming public policy priority. India's urban population to grew by 230 million from 1971 to 2008, and the country is on track to add another 250 million to its towns and cities in the next 20 years (Sankhe et al. 2010). India's capital, New Delhi, has received approximately 700,000 new residents every year (almost as many as the entire population of Leeds, United Kingdom) (Sankhe et al. 2010).

Apart from migration, urban centres witness fast movements of goods and people moving within as well as in and out of them. Cities account for 80% of global gross domestic product (GDP), and millions of people transit large airports and seaports in cities such as Mumbai, Manila and Nairobi (UN-Habitat 2011). Dynamism too has implications for resilience. On the one hand, a dynamic economic and demographic base provides a solid foundation for resourcing and unearthing innovative solutions for resilience. On the other hand, dynamism can also lead to cities being most affected by some hazards. For instance, the fast movement of people in and out of urban centres is one reason that they account for 90% of all reported COVID-19 infections around the world (UN 2020).

The fact that cities are dense, diverse and dynamic also leads to a degree of complexity. While complexity can pertain to social interaction or economic enterprise this issue is most pertinent in the context of urban governance. Complexity is a particularly useful concept for understanding urban politics and governance in cities of the Global South, where urban institutions are particularly "knotty" and the interaction between formal and informal institutions is not always clear (Beall et al. 2010: 5).

Urban centres across the Global South have multiple different formal and informal institutions with interacting and overlapping mandates, resulting in highly complex governance arrangements. This is routinely true for megacities, but even in small cities, it is not uncommon to find multiple government entities charged with planning and/or service delivery. It is also common to find both informal and formal actors delivering urban services such as water and waste management concurrently. Again, this has implications for resilience. The involvement of multiple government agencies and informal actors creates multiple options for service delivery in times of crisis, but this populated institutional landscape also leads to high coordination and transaction costs that can impede resilience (Bahadur and Tanner 2014).

1.3 THE URBAN RISK CONTEXT

These characteristics of the urban context interact with a variety of risks that are evolving in urban areas. A special report of the Intergovernmental Panel on Climate Change underlines how a number of climate-induced extreme events are on the rise (IPCC 2012). These include an increase in temperature extremes and the number of warm days and warm nights. For instance, in the United States the number of extremely hot days could go up by 300% by the end of the 21st century (under the high emissions scenario) (Center for Climate and Energy Solutions n.d.). The rise in temperature is accompanied by growing incidences of extreme rainfall events, manifested, for example, as a trebling of widespread extreme rain events in central India between 1950 and 2015 (Roxy et al. 2017). This rise in temperature also contributes to ocean expansion and the consequent exacerbation of coastal storm surge events (IPCC 2012).

The rise in events that are outside the historical experience means that risk reduction protocols can no longer rely solely on estimating the probability of extreme events using past patterns. A good illustration is Hurricane Katrina in 2005, which led to almost 2000 deaths and damaged property worth USD 125 billion. This was a "black swan" event; that is, it was considered so unlikely that protective infrastructure was not geared to deal with it (Nafday 2009). More specifically, flood control infrastructure around New Orleans (such as levees) was breached at more than 50 points, leaving 80% of the city underwater. While there were multiple causes of this disaster the city's infrastructure was designed for hazards predicted to occur once every 100 years; Hurricane Katrina lay outside that range and was considered a 400-year event. Apart from these hydrometeorological extreme events, other types of shocks are rising in frequency too. Statistically there is a 1% chance of a pandemic such as COVID-19 occurring in any given year, but due to the projected growth in global travel, urbanisation, environmental degradation and increased human–animal contact, such devastating outlier events are likely to become more frequent (GAVI 2020; Fan et al. 2018).

The concurrence of multiple different kinds of risk will lead to compound and complex crises. For example, in May 2020 an area of low pressure started to develop in the Bay of Bengal, and by the middle of that month this became Cyclone Amphan, a major super cyclonic storm event. A few days later, Amphan made landfall in the state of Odisha, India. Ordinarily, the state would be well prepared, having faced a number of storms in the past. However, as Amphan hit Odisha, the provincial government was busy battling COVID-19. A quarter of Odisha's cyclone shelters were being used as medical centres for those quarantined due to COVID-19, and a number of key government personnel had been deputed to battling the pandemic. 4.4 million people were affected by Amphan with preparedness activities and emergency services severely stretched.

Cities account for the majority of all reported COVID-19 infections around the world. As of 15 September 2020, 92 extreme weather events had unfolded in locations that were battling the pandemic, exposing 51.6 million people globally to an overlap of floods, droughts or storms and the pandemic (Walton and van Aalst 2020). It is no longer adequate for cites to have linear risk management plans for different crises. Governments, the private sector and citizens need to fundamentally transition to operating in an environment of "radical uncertainty" (King and Kay 2020). This entails the adoption of principles that lie at the core of the concept of resilience, such as flexibility, redundancy and modularity (Bahadur et al. 2013).

Another major evolution in the urban context is the growth of transboundary and teleconnected risk. Growth in the density of global economic, political and social networks means that perturbation in one part of the globe can easily lead to disruption in another. In the summer of 2011, Tropical Storm Nock-ten made landfall in Thailand, with extreme rainfall and consequent flooding across major cities, including Bangkok. Apart from the direct severe economic damage that these floods caused within cities, their impact reverberated across the world. This included a global spike in the price of computer hard disks and severe disruption to car production in Indonesia, Japan, Malaysia, North America, Pakistan, the Philippines, South Africa and Vietnam, as components for these products, sold all over the world, were manufactured in flood-affected areas (Avory et al. 2015; Fuller 2011). These production impacts in turn had far-reaching economic and social consequences. In another example in September of the same year, a transmission line tripped due to extreme heat in Yuma, Arizona, leading to a chain of events that culminated in the shutdown of the San Onofre nuclear power plant in California (Moser and Hart 2015). This in turn caused power shortages across US and Mexican cities and cascading impacts on other urban services, including sewage spills and water shortages. More recently, COVID-19-related shutdowns in China have threatened supply of essential goods to cities across sectors from textiles to pharmaceuticals. This geographical and temporal separation between hazard and impacts makes it particularly difficult to anticipate and plan for such events.

1.4 PIVOTS FOR TACKLING CLIMATE RISK IN CITIES OF THE GLOBAL SOUTH

This evolution in risk is rendering existing systems of risk management redundant, and urban resilience practices need to step up if they are to be effective. This book calls for a reset to take stock and examine new directions for understanding and acting on risk, arguing that the time is ripe for those in the world's cities to "pivot" towards new ways of enhancing resilience. We use the metaphor of a pivot as a central point on which a mechanism turns to indicate the turning point and redirection of existing practices. Chapters 2 to 6 of this book identify critical pivots, in five key areas of action, towards novel approaches to enhancing urban resilience. These pivots can take one of four principal forms.

First, a pivot can entail the use of new technologies to deliver greater impact. Chapter 2 explores how "big data" approaches can help overcome the challenges with existing ways of acquiring and analysing information on the risks that urban centres face. Such data, harnessed at great velocity and in large volumes with great variety, emanates from a range of high-tech data sources, helping overcome some of the existing challenges of measuring and tracking resilience (SAS 2020). This includes data on the use of mobile phones, automatic teller machines (ATMs), point-of-sales machines, remotely sensed and drone-mounted cameras, social media platforms, open mapping software and artificial intelligence (AI) applications, all of which can help determine the hazards, vulnerability and exposure that urban populations face.

Second, a pivot can entail new approaches for risk management in cities. For instance, Chapter 3 recommends a pivot away from tackling the "proximate" causes of risk to its structural drivers (i.e. issues of power, empowerment and agency). Chapter 5 suggests a pivot away from the emphasis on managing risk through infrastructure solutions towards building the capacity of those running critical urban systems to make decisions in conditions of uncertainty through approaches such as "adaptive management" and more politically smart understanding of gender, race, power and other structural drivers. Chapter 6 illustrates the benefits of a shift from focusing on streams of finance for building resilience that emanate from outside cities to more endogenous modes of financing risk management.

Third, the book proposes shifts in modes of engagement for delivering resilience. Implicit in Chapter 2 is a call for closer engagement between technology innovators and urban resilience professionals. Chapter 3 argues for deliberate engagement with actors and institutions at different scales of governance, as opposed to an exclusive focus on the local level. Chapter 4 outlines the critical importance of engaging with those living in informal settlements and operating within the informal economy as part of any process to build urban resilience. Chapter 6 explains how strengthening local capacity and private sector engagement is critical to securing the substantial resources needed for building resilience.

Finally, the pivot can be related to new ways of understanding and conceptualising risk and resilience. As the concept of resilience has entered the mainstream, so its use has been extensively critiqued (Tanner et al. 2017). Chapter 5 calls for a reorientation of urban resilience that emphasises the role of soft systems, and more transformative engagement with communities are discussed in Chapter 3. Chapter 4 demands a contextual understanding of the informal economy that dominates resilience processes in many urban areas, especially in the Global South. Throughout, the book makes the case for a normative underpinning to resilience thinking and practice based on rights, equity and justice (Ziervogel et al. 2017).

We contend in this book that such pivots towards new technology, management approaches, modes of engagement and concepts are needed to reset resilience practices in urban centres so that the lives of billions of urban residents across the Global South not only function but flourish despite multiple interacting shocks and stresses.

REFERENCES

Avory, B., Cameron, E., Rickson, C. and Fresia, P. (2015). *Climate Resilience and the Role of the Private Sector in Thailand.* Hong Kong: BSR.

Bahadur, A. (2014). *Policy Climates and Climate Policies: Analysing the Politics of Building Resilience to Climate Change* [PhD thesis]. University of Sussex, UK.

Bahadur, A. V., Ibrahim, M. and Tanner, T. (2013). Characterising resilience: Unpacking the concept for tackling climate change and development. *Climate and Development*, *5*(1), 55–65.

Bahadur, A., Pichon, F. and Tanner, T. (2016). *Enhancing Urban Resilience: Seven Entry Points for Action.* Manila: Asian Development Bank.

Bahadur, A. V. and Tanner, T. (2014). Policy climates and climate policies: Analysing the politics of building urban climate change resilience. *Urban Climate*, *7*, 20–32.

Bahadur, A. V. and Thornton, H. (2015). Analysing urban resilience: A reality check for a fledgling canon. *International Journal of Urban Sustainable Development*, *7*(2), 196–212.

Batty, M. (2008). *Cities as Complex Systems: Scaling, Interaction, Networks, Dynamics and Urban Morphologies*. UCL Centre for Advanced Spatial Systems Analysis Working Paper 131. London: University College London.

Beall, J., Basudeb, G. K. and Kanbur, R. (2010). *Urbanization and Development: Multidisciplinary Perspectives*. Oxford: Oxford University Press.

Brown, A., Dayal, A. and Rio, C. R. (2012). From practice to theory: Emerging lessons from Asia for building urban climate change resilience. *Environment and Urbanization*, *24*(2), 531–556. doi:10.1177/0956247812456490

Brown, A. and Kernaghan, S. (2011). Beyond climate-proofing: Taking an integrated approach to building climate resilience in Asian Cities. *UGEC Viewpoints*, *6*, 4–7.

Bull-Kamanga, L., Diagne, K., Lavell, A., Leon, E., Lerise, F., Macgregor, H. and Yitambe, A. (2003). From everyday hazards to disasters: The accumulation of risk in urban areas. *Environment and Urbanization*, *15*(1), 193–203. doi:10.1630/095624703101286457

Button, C., Mias-Mamonong, M. A., Barth, B. and Rigg, J. (2013). Vulnerability and resilience to climate change in Sorsogon City, the Philippines: Learning from an ordinary city? *Local Environment*, *18*(6), 705–722. doi:10.1080/13549839.2013.798632

Buxton, M., Carey, R. and Phelan, K. (2016). The role of peri-urban land use planning in resilient urban agriculture: A case study of Melbourne, Australia. In B. Maheshwari, V. P. Singh and B. Thoradeniya (Eds), *Balanced Urban Development: Options and Strategies for Liveable Cities*, Water Science and Technology Library, Vol. 72 (pp. 153–170). Cham: Springer. doi:10.1007/978-3-319-28112-4

Carmin, J., Anguelovski, I. and Roberts, D. (2012). Urban climate adaptation in the Global South: Planning in an emerging policy domain. *Journal of Planning Education and Research*, *32*(1), 18–32.

Center for Climate and Energy Solutions. (n.d.). *Heat waves and climate change*. Retrieved 23 November, 2020, from https://www.c2es.org/content/heat-waves-and-climate-change/

Dodman, D. and Carmin, J. (2011). *Urban Adaptation Planning: The Use and Limits of Climate Science*. London: International Institute of Environment and Development.

Ellis, P. and Roberts, M. (2016). *Leveraging Urbanization in South Asia: Managing Spatial Transformation for Prosperity and Livability*. Washington DC: The World Bank.

Ernstson, H., van der Leeuw, S. E., Redman, C. L., Meffert, D. J., Davis, G., Alfsen, C. and Elmqvist, T. (2010). Urban transitions: On urban resilience and human-dominated ecosystems. *Ambio*, *39*(8), 531–545.

Evans, J. P. (2011). Resilience, ecology and adaptation in the experimental city. *Transactions of the Institute of British Geographers*, *36*(2), 223–237.

Fan, V. Y., Jamison, D. T. and Summers, L. H. (2018). Pandemic risk: How large are the expected losses? *Bulletin of the World Health Organization*, *96*(2), 129–134.

Folke, C. (2006). Resilience: The emergence of a perspective for social "ecological systems analyses". *Global Environmental Change*, *16*(3), 253–267.

Friend, R. and Moench, M. (2013). What is the purpose of urban climate resilience? Implications for addressing poverty and vulnerability. *Urban Climate*, *6*, 98–113. doi:10.1016/j.uclim.2013.09.002

Fuller, T. (2011, 6 November). *Thailand flooding cripples hard-drive suppliers*. The New York Times. Retrieved 23 November, 2020, from https://www.nytimes.com/2011/11/07/business/global/07iht-floods07.html

Gasper, R., Blohm, A. and Ruth, M. (2011). Social and economic impacts of climate change on the urban environment. *Current Opinion in Environmental Sustainability*, *3*(3), 150–157.

GAVI. (2020). *5 reasons why pandemics like COVID-19 are becoming more likely*. Retrieved 23 November, 2020, from https://www.gavi.org/vaccineswork/5-reasons-why-pandemics-like-covid-19-are-becoming-more-likely

Gencer, E., Folorunsho, R., Linkin, M., Wang, X., Natenzon, C. E., Wajih, S., Mani, N., Esquivel, M., Ali Ibrahim, S., Tsuneki, H., Castro, R., Leone, M., Panjwani, D., Romero-Lankao, P. and Solecki, W. (2018). Disasters and risk in cities. In C. Rosenzweig, W. Solecki, P. Romero-Lankao, S. Mehrotra, S. Dhakal and S. Ali Ibrahim (Eds), *Climate Change and Cities: Second Assessment Report of the Urban Climate Change Research Network* (pp. 61–98). New York: Cambridge University Press.

Godschalk, D. R. (2003). Urban hazard mitigation: Creating resilient cities. *Natural Hazards Review*, *4*(3), 136–143.

Google Trends. (2020). Trend search using keyword "resilience". Retrieved 23 November, 2020, from https://trends.google.com/trends/explore?date=all&q=resilience

Government of India. (2011). *2011 Census data*. Office of the Registrar General & Census Commissioner. Retrieved 23 November, 2020, from https://censusindia.gov.in/2011-Common/CensusData2011.html

Gunderson, L. H. and Holling, C. S. (2001). *Panarchy: Understanding Transformations in Human and Natural Systems.* Washington DC: Island Press.

Güneralp, B., Güneralp, İ. and Liu, Y. (2015). Changing global patterns of urban exposure to flood and drought hazards. *Global Environmental Change, 31*, 217–225.

Hedblom, M., Andersson, E. and Borgström, S. (2017). Flexible land-use and undefined governance: From threats to potentials in peri-urban landscape planning. *Land Use Policy, 63*, 523–527.

Holling, C. S. (1973). Resilience and stability of ecological systems. *Annual Review of Ecology and Systematics, 4*(1), 1–23.

Hoon, C. Y. (2006). Assimilation, multiculturalism, hybridity: The dilemmas of the ethnic Chinese in post-Suharto Indonesia. *Asian Ethnicity, 7*(2), 149–166.

Intergovernmental Panel on Climate Change. (2012). *Managing the Risks of Extreme Events and Disasters to Advance Climate Change Adaptation. A Special Report of Working Groups I and II of the Intergovernmental Panel on Climate Change* (C. B. Field, V. Barros, T. F. Stocker, D. Qin, D. J. Dokken, K. L. Ebi, M. D. Mastrandrea, K. J. Mach, G.-K. Plattner, S. K. Allen, M. Tignor and P. M. Midgley, Eds). Cambridge, UK: Cambridge University Press.

King, M. and Kay, J. (2020). *Radical Uncertainty*. London: The Bridge Street Press.

Korf, B. (2002). *Does PRA Make Sense in Democratic Societies?* London: International Institute of Environment and Development.

Leichenko, R. (2011). Climate change and urban resilience. *Current Opinion in Environmental Sustainability, 3*(3): 164–168.

Mani, N. (2015). *Farming on the city's periphery to enhance resilience*. International Institute for Environment and Development. Retrieved 23 November, 2020, from https://www.iied.org/farming-citys-periphery-enhance-resilience

Meerow, S., Newell, J. P. and Stults, M. (2016). Defining urban resilience: A review. *Landscape and Urban Planning, 147*, 38–49.

Moser, S. C. and Hart, J. A. F. (2015). The long arm of climate change: Societal teleconnections and the future of climate change impacts studies. *Climatic Change, 129*(1–2), 13–26.

Nafday, A. M. (2009). Strategies for managing the consequences of black swan events. *Leadership and Management in Engineering, 9*(4), 191–197.

Pelling, M. (2011). Urban governance and disaster risk reduction in the Caribbean: The experiences of Oxfam GB. *Environment and Urbanization, 23*(2), 383–400.

Pelling, M. and Manuel-Navarrete, D. (2011). From resilience to transformation: The adaptive cycle in two Mexican urban centers. *Ecology and Society*, *16*(2), Article 11.

Prashar, S., Shaw, R. and Takeuchi, Y. (2012). Community action planning in East Delhi: A participatory approach to build urban disaster resilience. *Mitigation and Adaptation Strategies for Global Change*, *18*(4), 429–448.

Ramalingam, B. and Jones, H. (2008). *Exploring the Science of Complexity: Ideas and Implications for Development and Humanitarian Efforts*. London: Overseas Development Institute.

Revi, A., Satterthwaite, D. E., Aragón-Durand, F., Corfee-Morlot, J., Kiunsi, R. B. R., Pelling, M., Roberts, D. C. and Solecki, W. (2014). Urban areas. In V. R. Barros, C. B. Field, D. J. Dokken, M. D. Mastrandrea, K. J. Mach, T. E. Bilir, M. Chatterjee, K. L. Ebi, Y. O. Estrada, R. C. Genova, B. Girma, E. S. Kissel, A. N. Levy, S. MacCracken, P. R. Mastrandrea and L. L. White (Eds), *Climate Change 2014: Impacts, Adaptation, and Vulnerability. Part A: Global and Sectoral Aspects. Working Group II Contribution to the Fifth Assessment Report of the Intergovernmental Panel on Climate Change* (pp. 535–612). New York: Cambridge University Press.

Rockefeller Foundation. (2009). *Building Climate Change Resilience*. New York: Rockefeller Foundation.

Rockefeller Foundation and Arup. (2015). City Resilience Framework. https://www.rockefellerfoundation.org/wp-content/uploads/City-Resilience-Framework-2015.pdf

Rosenzweig, C. and Solecki, W. (2018). *The Future We Don't Want*. New York: Urban Climate Change Research Network.

Roxy, M. K., Ghosh, S., Pathak, A., Athulya, R., Mujumdar, M., Murtugudde, R., Terray, P. and Rajeevan, M. (2017). A threefold rise in widespread extreme rain events over central India. *Nature Communications*, *8*(1), 1–11.

Sankhe, S., Vittal, I., Dobbs, R., Mohan, A., Gulati, A., Ablett, J., Gupta, S., Kim, A., Paul, S., Sanghvi, A. and Setyy, G. (2010). *India's Urban Awakening: Building Inclusive Cities, Sustaining Economic Growth*. New Delhi: McKinsey Global Institute.

SAS. (2020). *Big data: What is it and why it matters*. SAS Insights. Retrieved 23 November, 2020, from https://www.sas.com/en_us/insights/big-data/what-is-big-data.html

Satterthwaite, D. (2013). The political underpinnings of cities' accumulated resilience to climate change. *Environment and Urbanization*, *25*(2), 381–391.

Satterthwaite, D. and Dodman, D. (2013). Towards resilience and transformation for cities within a finite planet. *Environment and Urbanisation*, *25*(2), 291–298.

Schipper, E. L. F., Eriksen, S. E., Fernandez Carril, L. R., Glavovic, B. C. and Shawoo, Z. (2020). Turbulent transformation: Abrupt societal disruption and climate resilient development. *Climate and Development*. Advance online publication. https://doi.org/10.1080/17565529.2020.1799738

Silva, J., Kernaghan, S. and Luque, A. (2012). A systems approach to meeting the challenges of urban climate change. *International Journal of Urban Sustainable Development*, *4*(2), 125–145.

Simonsen, S. H. (2007). *Resilience dictionary*. Stockholm Resilience Centre. Retrieved 23 November, 2020, from www.stockholmresilience.org/research/whatisresilience/resiliencedictionary.4.aeea46911a3127427980004355.html

Simonsen, S. H., Biggs, R., Schluter, M., Schoon, M., Bohensky, E., Cundill, G., Dakos, V., Daw, T., Kotschy, K., Leitch, A., Quinlan, A., Peterson, G. and Moberg, F. (2015). *Applying Resilience Thinking*. Stockholm: Stockholm Resilience Centre.

Swanstrom, T. (2008). *Resilience: A Critical Examination of the Ecological Framework*. Working Paper 2008-7. Oakland, CA: University of California Berkley, Institute of Urban and Regional Development.

Tanner, T., Bahadur, A. and Moench, M. (2017). *Challenges for Resilience Policy and Practice*. London: Overseas Development Institute.

Tyler, S. and Moench, M. (2012). A framework for urban climate resilience. *Climate and Development*, *4*(4), 311–326.

UN-Habitat. (2011). *The Economic Role of Cities*. Nairobi: United Nations Human Settlements Programme.

United Nations. (2020). *Policy Brief: COVID-19 in an Urban World*. New York: United Nations.

United Nations Department of Economic and Social Affairs. (2018). *World Urbanization Prospects*. New York: United Nations Department of Economic and Social Affairs.

Walton, D. and van Aalst, M. (2020). *Climate-Related Extreme Weather Events and COVID-19*. Geneva: International Federation of Red Cross and Red Crescent Societies.

Wardekker, J. A., de Jong, A., Knoop, J. M. and van der Sluijs, J. P. (2010). Operationalising a resilience approach to adapting an urban delta to uncertain climate changes. *Technological Forecasting and Social Change*, *77*(6), 987–998.

Wong, T. E., Bakker, A. M. and Keller, K. (2017). Impacts of Antarctic fast dynamics on sea-level projections and coastal flood defense. *Climatic Change*, *144*, 347–364.

World Bank. (2020). *Data: Urban population growth (annual %) - low income*. Retrieved 23 November, 2020, from https://data.worldbank.org/indicator/SP.URB.GROW?locations=XM

Ziervogel, G., Pelling, M., Cartwright, A., Chu, E., Deshpande, T., Harris, L., Hyams, K., Kaunda, J., Klaus, B., Michael, K., Pasquini, L., Pharoah, R., Rodina, L., Scott, D. and Zwe, P. (2017). Inserting rights and justice into urban resilience: A focus on everyday risk. *Environment and Urbanization*, *29*(1), 123–138.

Chapter 2

Data for urban resilience: From mainstream to innovative approaches

THE foundation of initiatives to build the climate resilience of towns and cities lies in acquiring and analysing information and data. Urban areas have innate advantages in this regard, as these are the locations where intellectual and technological capital is concentrated. Building resilience requires an acute understanding of interdependent urban sectors, including water and sanitation, electricity, housing, natural resource management, transportation, disaster risk management and delivery of basic services. Significant data on these sectors is collected as part of "business as usual" planning and operations in urban areas (Bahadur et al. 2016). Even smaller urban agglomerations in low income country settings generally have units within government that gather information on basic services as well as shocks and stresses. As a result, in most places, a foundation exists for gathering and analysing risk data and information.

The concept of risk forms the guiding architecture for this chapter, which examines the dominant approaches that have been used to collect data and information across the elements of risk and the challenges these present. The chapter then makes a case for pivoting towards a new generation of innovative "big data" approaches that rely on information and communication technologies to help overcome these challenges.

2.1 MAINSTREAM DATA COLLECTION AND ANALYSIS

There are extensive data and information needs for enhancing the resilience of towns and cities to shocks and stresses. We organise the review of methods for collecting and analysing data using the three subcomponents of a risk framework, namely hazards, exposure and vulnerability.

2.1.1 Hazards

A hazard is "the potential occurrence of a natural or human-induced physical event that may cause loss of life, injury, or other health impacts, as well as damage and loss to property, infrastructure, livelihoods, service provision, and environmental resources" (IPCC 2014: 5). Hazards may emanate from factors that are natural or human-induced, but this analysis largely limits itself to those that are hydrometeorological and prone to influence by climate change. Hazards can become "disasters" when they meet assets and communities that are exposed and vulnerable. Urban areas face multiple hazards; a survey of European cities found that 92.5% of cities reported facing climate-change-induced impacts and hazards (Groth et al. 2016).

Weather forecasts are used widely to inform climate resilience-building processes. These forecasts typically draw on recent monitoring data and use variables such as air pressure, humidity, cloud formation and current temperature to forecast hydrometeorological conditions over the short term (Lynch 2010). Such data are collected by instruments on the ground, on balloons and sometimes from satellites as well. The data inform models to make predictions that can be used to forecast conditions (Lynch 2010). These forecasts are useful in understanding climate variability and, as part of this, more immediate hazards. This means that at times they may lead to resilience measures that are effective in the face of short-term risks but less prepared for changes that occur over an extended period (usually decades or longer) (WMA 2019).

Data for longer-term climate change is normally generated through atmosphere–ocean general circulation models, more commonly known as general circulation models (GCMs) (Herron et al. 2014). GCMs depict the climate using a three-dimensional grid over the globe, typically having a horizontal resolution of between 250 and 600 kilometres (IPCC 2019). These are "physics-based mathematical representations of the Earth's climate system over time that can be used to estimate the sensitivity of the climate system to changes in atmospheric concentrations of greenhouse gases (GHGs) and aerosols" (Bader et al. 2018: 48). Regional climate models (RCMs) employ a similar method but simulate changes at a much lower spatial scale, generally in the order of 25–50 kilometres. Both types of model can generate future projections for understanding changes such as those in temperature, sea level rise and precipitation under different climate change scenarios. Results can feed concomitant "impact models" that, for instance, may determine the impact of changes in temperature to the spread of disease or changes in crop yields. These models add on to an understanding of climate variability emanating from weather forecasts.

Satellite remote sensing (SRS) is another important method for collecting data across a range of hazards (Kaku 2019). For floods, floodplain delineation maps, historical data and soil cover and soil moisture data obtained through SRS imagery have been used successfully to help manage risk (Tralli et al. 2005). For landslides, slope maps, slope stability, elevation geology, soil type, areas of standing water and land use data derived from satellite imaging can be very helpful in informing resilience planning (Tralli et al. 2005). SRS has been used to detect landslides in urban areas in a number of ways, including through the use of infrared scanners to determine seepage areas that lubricate the movement of rocks and soil by ascertaining maximum temperature difference between the terrain and the effluent groundwater (Guzzetti et al. 2012; De Blasio 2011).

Similarly, for hurricanes, SRS can be used to detect cloud observations, temperature observations and the path of hurricanes that have impacted particular areas (Ruf et al. 2013). SRS can also be used for monitoring of stresses; for example, monitoring creeping conditions of drought and desertification using data on soil moisture content, crop conditions and natural vegetation (Wang et al. 2015). Early warning systems for a variety of hazards draw on SRS data (sometimes as part of a suite of data streams) to trigger action to manage risk and build resilience. For example, SRS data can support comparison of the current vegetation index with long-term averages to trigger warnings for water scarcity, drought, food security and heatwaves (Letouzé et al. 2015). SRS data can also be useful in determining the likelihood of a whole range of hazards. Different techniques can be used to detect various hazards, and the proficiency of those interpreting data is vital.

There has been more limited use of approaches for systematically gathering and analysing people's perceptions of hazards that impact them. This is important as research suggests that perceptions of decision makers are a primary determinant of the scope of urban adaptation agendas (Lee and Hughes 2017). Such exercises can also feed and validate climate models as well as filling gaps left by climate modelling and SRS analysis. In an analysis of perceptions data from Nepal, Tanner and colleagues (2018: 1) found that "changes in windstorms are a critical impact factor for agricultural livelihoods and hence for adaptation needs, in contrast with the traditional focus in climate change modelling on rainfall and temperature data".

2.1.2 Exposure

Understanding the degree to which urban areas are exposed to these hazards is crucial to determining how to enhance resilience. Exposure is understood as the "the presence of people, livelihoods, species or ecosystems, environmental functions, services, and resources, infrastructure, or economic, social, or cultural assets in places and settings that could be adversely affected" (IPCC 2014: 5). Such exposure is acute in the many towns and cities located along coasts and rivers – areas that are at high risk from hydrometeorological hazards such as cyclones, coastal storm surges and floods (Lankao and Qin 2011). Currently, 65% of the world's urban population live in coastal zones that are highly exposed to a range of hydrometeorological hazards, and this proportion is likely to increase to 74% by 2025 (Gencer et al. 2018).

Within the city, exposure may be greater in some areas than others, such as in informal settlements on floodplains or unstable hill slopes. Low-lying coastal zones also present increased exposure; 14 million urban Indians live in areas where the elevation is 5 metres below sea level (CIESIN 2013). Factors determining the exposure of assets and people include land use, land availability, in-migration, population size, city size, population density, quality of infrastructure, type and quality of the built environment and its regulation, and governance structures (Mehrotra et al. 2009; Dickson et al. 2012; GIZ 2014). Exposure data to inform resilience-building must also consider the potential knock-on effects and interlinkages; the exposure of certain critical systems may have graver implications for urban resilience than others (Connelly et al. 2018).

SRS and geographic information system (GIS) data have been used extensively to understand how people and assets might be affected by shocks and stresses induced by climate change. SRS can gather information on many parameters used in measuring exposure. This includes topography and elevation that show the height and shape of mountains, depth of the ocean bottom, steepness of slopes, physical contours and other such land features (USGS 2019). These have an important bearing on the exposure of assets within the land in question (GIZ 2014). For example, topographic data from Gorakhpur city in India reveals that the city is "bowl shaped" and even minor spikes in precipitation leaves the city's population exposed to flooding (Bahadur 2014). Geological and geomorphological data can also gauge the degree of exposure through data on landforms and their processes, form and sediments at the surface of the earth (Tooth and Viles 2014). These, in turn, can influence the way hazards might cause losses. For instance, studies have demonstrated how a thicker soil layer controls run-off more effectively and buffers against flooding. Populations living on or near areas where soil is thin can therefore be more exposed to flood risk (Naylor et al. 2017).

Physical parameters collected by SRS data can also help determine which populations and assets are exposed to climate impacts. For instance, one study correlated satellite imagery of night-time lights with flood damage data from the Emergency Events Database (established by the Centre for Research on the Epidemiology of Disasters) across major global river networks between 1991 and 2012. It found that the increasing density of night-time lights was positively correlated with economic damage per unit area (USD/km^2) (Ceola et al. 2014). This retrospective analysis demonstrates how acquiring population density estimates using SRS data with other variables facilitates the computation of levels of exposure to various hazards. Similarly, SRS data can be used to determine the function, quality and density of infrastructure in urban centres. This, in turn, can be used to calculate economic loss exposure using a variety of analytical techniques, such as "damage curves" that help understand the degree of damage that occurs to structures and contents for varying degrees of hazard severity (USACE 2006).

Census data is another important information source for determining exposure. For instance, the Census of India identifies numbers of people and assets within particular areas (Government of India 2011). It categorises houses as "good", "liveable" or "dilapidated" and provides rich details on: materials of roofs, walls and floors; number of rooms; type of latrine; source of lighting; kitchen facilities; information, communication and telecommunication devices; and vehicles. Together, this provides a picture of *what* is exposed, and it can be combined with other parameters to provide a more comprehensive picture of *how* this exposure occurs to compute the potential for loss. Most urban centres will also have a historical record of the shocks and stresses they have experienced and the parts of the city that were most affected, although it is increasingly acknowledged that exclusive reliance on historical record is not adequate to determine potential exposure to future shocks and stresses (Bader et al. 2018).

Finally, participatory methods can be employed to gauge exposure. Participatory maps and transect walks, for example, typically entail the development of criterion for observation (e.g. low-lying areas that might be exposed to inundation or urban canyons that might be more exposed to extreme temperatures) via walks through a contained geographic space with people that are knowledgeable about the area to create a simple model or transect diagram of exposed areas (Moser and Stein 2011). Exercises such as these have been undertaken the world over. One example comes from the Masantol municipality in the Philippines, where local communities in collaboration with a team of facilitators prepared a participatory three-dimensional map of the local area (see Box 2.1).

2.1.3 Vulnerability

Data and information are also needed on climate vulnerability, which describes the innate characteristics of a town and city that make it susceptible to the climate-related impacts and can be defined as "the propensity of exposed elements such as human beings, their livelihoods, and assets to suffer adverse effects when impacted by hazard events" (Cardona et al. 2012: 69). This is a composite of two subsidiary concepts: sensitivity and adaptive capacity. The former is "the degree to which a system will respond to a given change in climate, including beneficial and harmful effects" and the latter is "the ability of a system to adjust to climate

BOX 2.1 Participatory three-dimensional map in Masantol, Philippines

Participatory three-dimensional mapping involves the creation of relief maps using locally available materials, such as cartons and paper, which are overlapped with thematic layers of geographical information. Participants can use pins, yarns or small flags to depict landmarks and topographical features. In Masantol in the Philippines, local residents, with help from a team of facilitators, marked schools, churches, stores, village meeting centres, health centres, flood control facilities, mangroves, fish ponds, fishing grounds, roads and trail networks. They then identified the location of people with a heightened vulnerability to flood risk, such as the disabled, those with long-term illnesses, the elderly, young children and pregnant women. They also plotted resources and assets that can reduce risk, such as boats and vehicles. This exercise provided locally relevant insights on risk resilience, such as hotspots with concentrations of people vulnerable to flooding, and led to this zone being designated as a priority area for resilience building.

Source: Adapted from Cadag and Gaillard (2011)

change, including climate variability and extremes, to moderate potential damages, to take advantage of opportunities, or to cope with the consequences" (McCarthy et al. 2001: 89, 21). There is a vast literature on vulnerability and its determinants, but elements can be usefully categorised in terms of assets that influence how a system responds to climate impacts and the offsets their damaging consequences, including: human (e.g. levels of health and education), social (e.g. strength of networks and relationships), physical (e.g. quality and amount of protective infrastructure), natural (e.g. quality and distribution of ecosystem services) and financial (e.g. financial safety nets and savings) (Wang et al. 2015; Lankao and Qin 2011).

Just as urban centres across the world are highly exposed to the impact of climate change, they also suffer from a high degree of vulnerability. As noted in Chapter 1, the IPCC estimates that 1 billion urban dwellers live in contexts characterised by a high degree of vulnerability and low adaptive capacity (Revi et al. 2014).

Methods of gathering data on vulnerability are diverse, but often involve the preparation of vulnerability indices that contains various weighted indicators spanning human, social, physical, natural and financial assets. SRS can sometimes be used to identify the presence of access to these assets. One vulnerability study of Tegucigalpa city, Honduras, used empirical evidence to demonstrate that the socio-economic status of a household is indicative of their vulnerability. By identifying the proportion of built-up area relative to vegetated area, the road conditions in a neighbourhood, roof types, the type and quantity of physical infrastructure as well as the slope on which the area is located, it was possible to determine the status and, by extension, the vulnerability of an urban population using satellite imagery (Ebert et al. 2009). Other studies have used different metrics and indicators to identify vulnerability of populations across the world.

Vulnerability can also be determined using publicly available data on a variety of socio-economic parameters. Survey data (including from censuses) on indicators such as wealth, social networks, access to ecosystem services and access to education and healthcare can be combined to determine the vulnerability of a particular group of people. Where this data does not already exist, bespoke surveys can help. One study of the US Northeast Megaregion (focusing on large cities in the region) used neighbourhood-level social, economic and demographic census data to analyse vulnerability (Cox et al. 2006). This developed six indicators that include material resources (e.g. levels of poverty and dependence on social welfare); the built environment (e.g. housing density); access to information (e.g. language abilities and literacy levels); presence of children and persons with disabilities; numbers of elderly and elderly living alone; and race (because evidence points to high levels of vulnerability among African-American individuals).

In comparison to processes of identifying hazards and exposure, vulnerability assessment has benefited from a much wider use of participatory methods. The use of participatory vulnerability assessment (PVA) has become widespread as "a systematic process that involves communities and other stakeholders in an in-depth examination of their vulnerability" through the use of participatory techniques (Chiwaka and Yates 2012: 11). PVA methodologies broadly have a uniform set of steps, beginning with a situation analysis of vulnerability aimed at determining the demographic profile of the community, the constituencies that need to be included in the assessment, the prevalence and extent of vulnerability as well as the coping capacity of community members. This can be done through focus group discussions, timeline exercises, vulnerability mapping, developing seasonal calendars and undertaking livelihood analysis (Flower et al. 2018). Following this, PVA often focuses on tackling underlying causes or drivers of the vulnerabilities identified and prioritising drivers using a variety of techniques that could include cause–effect trees and concept mapping (Turnbull and Turvill 2012). Once this stage is complete, the team undertaking PVAs can proceed to establish the existing strategies, resources and assets used to reduce vulnerability and identify gaps (IFRC 1999). The final step entails the participatory consolidation of pathways to fill gaps in capacity.

2.2 KEY CHALLENGES FOR DATA COLLECTION AND ANALYSIS

The prevailing methods of collecting and analysing data and information on hazards, exposure and vulnerability discussed above face a range of challenges. These limitations can be grouped according to three broad issues. First, uncertainties in climate data and information; second, limitations to the degree with which data and information can describe changes at particular points in time and at fine spatial scales; third, challenges with the reliability and veracity of the data. Together, these three limitations have contributed to a considerable gap between the acquisition of data and information and its use to inform decision-making. Overcoming these challenges is essential to ensure climate data and information is used more effectively in decision-making.

2.2.1 *Certainty*

There is now wide agreement that projections emanating from climate models are steadily improving, but uncertainty remains (Scoones 2019; Bader et al. 2008). This is because of "cascading uncertainty" at the different stages of model preparation (Maslin 2013) (see Box 2.2).

Such uncertainties in turn limit the utility of data provided to decision makers. Studying the uptake of climate information in Zambia and Malawi, Jones and colleagues (2015: 813) suggest that the current practice of disseminating information on changes in annual average temperature or precipitation "is of little practical use" and that "information on decision-relevant events such as changes to the onset of the rainy season, frequency and duration of dry spells early in the growing season,

BOX 2.2 Cascading uncertainty in climate modelling

GCMs/RCMs rely on data on future GHG emissions, which in turn are based on estimates of future economic growth (Maslin 2013). While scientists and researchers attempt to accommodate a plethora of environmental, economic and demographic variables that influence economic development, they are unable to include all the factors that might have a bearing on the complex processes that underpin this, contributing to a degree of uncertainty in the scenarios of GHG emissions that are fed into climate models.

GCMs/RCMs model the impact that the emission of GHGs will have on climate change, but are unable to completely accommodate the full breadth of land–air–water interactions necessary for delivering a foolproof projection (Maslin 2013). For instance, cloud formation is one such interaction that is not adequately understood and accommodated by models (Gentine et al. 2018). This is because high-density clouds can lead to an enhanced albedo effect and have a negative impact on global warming, whereas low-density clouds can contribute to the greenhouse effect, positively impacting temperature rise.

Moreover, projections from climate models are used to model hazards and impacts on various sectors (e.g. how temperature rise impacts crop loss) (Harrison et al. 2015). These include additional variables to make projections, thus adding another layer of uncertainty. This in no way means that climate and impact models are not at all useful; it is just that certain projections emanating from them are presented within wide ranges to account for this uncertainty. Over the years they have provided incontrovertible evidence of the link between GHG emissions and global temperature rise, and they have given "an insight into the possible climate of the future and clear choices about what future we would like to have" (Maslin 2013: 267).

or water availability for irrigation has far greater significance to local and national decision-makers".

This apart, the physical characteristics of cities lead to peculiarities in their climates that is poorly represented by commonly used climate change models. This is due to urban microclimates that influence precipitation, moisture and wind effects and the "urban heat island" effect, caused by the absorption of radiation, reduction of evapotranspiration due to impervious surfaces and release of heat from urban cooling and energy systems (Bader et al. 2018). High rates of urban expansion, especially in developing countries, means that the way the urban built environment and urban microclimates interact is constantly in flux. Finally, the heterogeneity in urban land use and the built environment poses specific challenges for impact modelling. The fact that urban form and morphology can vary block by block in a city means that changes in the climate can have vastly different impacts in different localities of a city (e.g. extreme rainfall may lead to floods in some neighbourhoods but not in others). This adds another layer of complication for those attempting to precisely model changes in climate and the impacts of this in urban areas.

2.2.2 Granularity

Closely associated with the challenge of uncertainty is the fact that the dominant methods for collecting and analysing climate information and data also lack the "granularity" needed for decision-making. More specifically, information and data from GCMs on changes in climate variables are projected at high spatial scales (in grids of 300 km^2), and these cannot be downscaled accurately. This is because the micro factors that determine, as an example, temperature in small areas cannot all be accurately included in the most widely used models (Piccolella 2013). Output at these scales may not be meaningful for urban planning that engages with relatively contained geographic units that are at times even finer than the scales at which projections from RCMs are delivered (i.e. 50 km^2) (Bader et al. 2018; van Oldenborgh et al. 2013; Randall et al. 2007).

Similarly, with SRS data there are trade-offs between spatial and temporal aspects. Satellites are better able to capture data (on hazards, exposure or vulnerability) at broader scales for longer durations and are generally only able to capture finer scales (crucial for urban climate change resilience initiatives) for shorter periods of time. Cloud cover can also inhibit the collection of certain kinds of data at fine scales (e.g. at the municipal level) (Adiningsih 2010). High-resolution SRS data are particularly important in urban contexts; for instance, the density of settlement patterns (especially in informal settlements) can make it difficult to ascertain the physical features of buildings (e.g. quality of construction) and the underlying infrastructure (e.g. drainage), which in turn makes it difficult to ascertain scale and nature of exposure.

Traditional data sources may also be less able to pick up important differences required to support our understanding of how vulnerability, risk and resilience vary across different groups. Such granularity is required at a basic level to understand and target factors such as the gendered nature of impacts and vulnerability but also the intersectionality of gendered power relations that mediate the resilience of different groups and places (Djoudi et al. 2016; Eastin 2018). Census data can provide granular data on vulnerability and exposure, but there is substantial variation in quality and resolution of census data, particularly in poorer countries or those with a history of conflict. For example, in the Democratic Republic of Congo, census data is available only at the level of territories (i.e. administrative units of 12,466 km^2 on average) (Deville et al. 2014). Data is also fixed to one point in time and, with most countries performing censuses only once a decade, quickly becomes outdated. This challenge is a particular problem in urban areas due to high rates of in- and out-migration (Willows 2003).

2.2.3 Veracity

The third set of limitations on dominant approaches to collecting and analysing climate information and data pertain to their veracity. Participatory approaches, used widely for collecting data on vulnerability and exposure, have been particularly criticised on this count. Some see these as reductionist where attempts are made to compress complex realities and information into simple formats (e.g. diagrams and crude maps), leading to a loss of vital detail (Brown et al. 2002). Others have argued that many participatory approaches do not employ robust sampling methods and, therefore, the data emanating from these methods are not representative; that is, they cannot furnish principles that are generalisable beyond the very limited contexts in which they unfold (Leurs 1996; GIZ 2014). It has been argued that at best these methods can estimate current vulnerabilities and exposure, but resulting data are not useful for forecasting or projecting future trends/scenarios (GIZ 2014). Also, due to the high amount of migration and other flux in urban areas, data collected through these methods are very quickly out of date. Others have highlighted that the lack of rigorous, established protocols means that participatory methods can be manipulated easily, and the views and priorities of those running such exercises can bias any data that are collected (Brown et al. 2002; Bahadur 2014; Hickey and Mohan 2004).

This apart, the success of participatory approaches is predicated on the need to include the full fabric of the community being studied. Otherwise, they risk privileging the views of the select few and marginalising other voices, which in turn can lead to the exclusion of the latter from the benefits of resilience-building initiatives. One of the important ways this can happen is through the "elite capture" of these participatory spaces. This process allows those with a greater amount of power as a result of their socio-economic status, caste or class position to dominate in participatory spaces (Lund and Saito-Jensen 2013).

Most participatory methods for gathering data and information on vulnerability and hazards in urban contexts stem from an earlier generation of participatory appraisal approaches that were uniquely structured for the rural context. Participatory approaches rely on a degree of social cohesiveness and are founded on the idea of a homogenous "community" (Korf 2002). However, urban centres across developing countries are socially dynamic spaces with rapid movement of people, and this can lead to people from diverse ethnic, geographic and cultural backgrounds settling in particular neighbourhoods that are then treated, artificially, as communities by those conducting participatory data collection exercises (Ward 2018).

2.3 PIVOTING TO INNOVATIVE APPROACHES

Over the past decade and a half, there has been an increase in methods and mechanisms to collect and analyse climate data and information using novel information and communication technologies. This includes, but is not limited to, the use of big data: an umbrella term to describe approaches for collecting a great variety of data that is verifiable, arrives in large volumes and can be collected at speed (Raghupathi and Raghupathi 2014). Closely associated with big data approaches are those that employ machine learning and AI. These contribute to the growing understanding of the empirical benefits of using high-frequency data sets to understand and track resilience (Knippenberg et al. 2019).

This section provides examples of the ways in which data from innovative, emergent approaches can be used alongside dominant approaches to overcome the challenges explored in Section 2.2. This provides an opportunity to build a better understanding of hazards, vulnerability and exposures for enhancing climate change resilience. Due to the emergent nature of these approaches, some may not have been used in urban contexts, but there is nothing inherently that prevents this in the future.

2.3.1 Hazards

Section 2.1 argued that the dominant method of acquiring and understanding data and information on hazards uses climate models, SRS and weather forecasts. Due to challenges of accuracy, certainty and precision of such data, these are failing to speak to the practical questions for which urban decision makers need answers (Jones et al. 2015: 813). A new generation of approaches is helping overcome these challenges.

As highlighted in Box 2.2, the uncertainty in modelling is a result of the inability of the current set of models to appropriate the full breadth of complicated interactions between the ocean, land and atmosphere as typified by inaccuracies in understanding cloud behaviour (Herron et al. 2014). "Cloud Brain" is a recent example of the way AI is being used to overcome this problem (Gentine et al. 2018). This system, currently being tested, uses deep learning techniques based on AI to run thousands of precise "small-scale, short-term models that show clouds evolving over short time periods with grid boxes just a few miles wide" to allow the AI system to develop an instinctive understanding of how clouds work (Jones 2018). This learning on cloud behaviour has then been inserted into global climate models with positive results; for instance, the system has predicted extreme rainfall events with a greater degree of accuracy (Gentine et al. 2018). Even as Cloud Brain demonstrates the way new technology can be used to better understand complex natural processes and provide more reliable projections, it has also delivered a clearer picture of impacts at fine spatial scales. As such, it has immense potential to benefit those attempting to understand climate impacts in urban areas. This is just one example of how machine learning and AI is being leveraged to overcome limitations of traditional climate modelling and inform efforts towards urban climate change resilience.

While applications such as Cloud Brain are helping develop models that deliver results with a greater degree of certainty, there has been a proliferation of approaches that rely on machine learning, AI and big data to forecast shocks, stresses and extreme weather events with a high degree of precision (see Box 2.3).

As discussed in Section 2.2.1, urban areas have microclimates that can buck broader climate and weather patterns due to

BOX 2.3 Artificial intelligence for precise weather forecasting

The global weather forecasting models currently in use rely on underlying dynamics that are calibrated to the mid-latitudes. As a result of this, their reliability decreases for forecasting tropical weather (Letouzé et al. 2015). The Iska tropical weather forecasting platform attempts to overcome this by delivering highly precise, two-day weather forecasts tailored to the individual user down to a spatial resolution of 3 kilometres in tropical climates (Vallgren and Petrykowska 2016). It does this by running remote sensing data from multiple satellites and ground-based stations through a multicore supercomputer to produce forecasts that are twice as accurate as the global models currently in use. Iska also includes a self-learning AI system that predicts the trajectories of individual thunderstorms by merging the numerical forecast system with observational data from lightning detection, satellite imagery and weather radars (Letouzé et al. 2015). The information is delivered to individual users as text messages. An evaluation of this platform by GIZ (Deutsche Gesellschaft für Internationale Zusammenarbeit) found that 90% of the users thought the forecasts allowed them to make effective decisions to deal with impending weather phenomenon. Therefore, while this is not useful for long-term projections, it does overcome barriers of decision relevance, timeliness and granularity in current methods of collecting information on hazards (Vallgren and Petrykowska 2016).

geospatial peculiarities in towns and cities as a result of the built environment. The urban heat island effect has proved particularly difficult to model and forecast, as it requires a dense network of meteorological sensors or SRS data that have very high spatio-temporal resolution. One initiative has overcome this challenge by crowdsourcing air temperature data from mobile phones (Overeem et al. 2017). Mobile phones that use lithium-ion batteries constantly monitor battery temperature to ensure that they are not damaged by attempts to charge them when they are overheating. Using Opensignal, a widely used application for tracking factors that weaken signal strength (including battery function), the initiative collected many battery temperature readings from eight cities across the world (Overeem et al. 2013). These readings were calibrated with air temperature readings from other sources (Bushwick 2013). This allowed formulation of an algorithm that captured the relationship between battery and air temperature, which was further refined to also include other variables such as the effects of the phones' own insulation and the owners' body temperature (Bushwick 2013). This process has been tested at scale and is now proven to accurately translate battery temperature into air temperature (Overeem et al. 2013). This provides one pathway to collect high-frequency, real-time, granular data on urban air temperatures that, in turn, can be used to determine the occurrence of extreme temperatures in urban areas. Approaches such as these are particularly useful for the urban context due to the density of data collection nodes (in this case, mobile phones).

Overall, this new generation of approaches are helping improve the reliability of projections from climate models as well as supplying decision makers and citizens with precise and dependable information on shocks, stresses and extreme events.

2.3.2 Exposure

Section 2.1 outlined the way data from SRS, census and surveys as well as participatory methods can be used to determine the degree to which urban assets and people are exposed to the impacts of climate. Challenges outlined in Section 2.2 included potential spatio-temporal trade-offs in SRS data, the manner in which survey and census data is static and in many cases does not extend insights at fine spatial scales, and how participatory approaches are prone to delivering results that may not be generalisable or verifiable. Here again, a new generation of approaches is helping overcome the challenges through the innovative use of information technology to deliver granular data that is more reliable.

One important set of approaches meeting these challenges fall under the broad category of volunteer geographic information. An example is Dar Ramani Huria ('open map' in Swahili), a community-based mapping project in Dar es Salaam, Tanzania – one of Africa's fastest-growing cities and one that is highly exposed to catastrophic flooding (AfDB 2014). One major flood event in 2011 affected 50,000 people and led to the displacement of 10,000, and another major flood impacted the city just three years later in 2014 (Reliefweb 2011, 2014).Existing city maps are of limited value as 70% of the population lives in informal settlements where buildings, land use and infrastructure is dynamic and not completely discernible from satellite maps (Soesilo 2015). To remedy this, the initiative employed drones to develop high-resolution imagery (up to 5 cm) to create digital maps of neighbourhoods that have historically been flood-prone. Once this was complete, a team of volunteers went into these communities to physically tag infrastructure such as drainage, roads, houses, schools, trees, businesses and other important features to these maps (Ospina 2018). Following this, they added information gathered from the community using standardised participatory methods on areas that flood frequently, the households that have the least capacity to respond to flooding and patterns from drainage.

This led to the development of a granular, three-dimensional map of the most flood-prone areas of Dar es Salaam that can pinpoint how populations and assets are exposed to flood impacts. Crucially, this information facilitates the development of hydrological and flood inundation models that assimilate information on elevation/drainage and allow experts "to run scenarios that demonstrate what happens if a certain amount of rainfall occurs, a river bursts its banks, or infrastructure is constructed – like a drain" (World Bank 2016: 13). Combining high technology with data gathered from communities to create these maps allowed local authorities to determine the areas that need urgent investment, identify highly exposed areas and households to be prioritised in disaster preparedness plans, and make a robust business case for drainage infrastructure in particular locations (Soesilo 2015). This is an example of an approach that helps overcome the subjectivity of participatory mapping methods, providing a decision-relevant granular snapshot of exposed areas that facilitates probabilistic as well deterministic modelling on future risk. Crucially, this is also considered a relatively fast and cost-effective method of mapping exposure (Natty 2018). Similar initiatives are under way in a whole range of cities in Sri Lanka, Indonesia and India among others (Harvey et al. 2018).

A key challenge in determining exposure is understanding where urban populations are located. This is particularly difficult because urban areas have high rates of migration and are highly dynamic. To overcome this challenge, a new set of approaches are using mobile phone data and call detail records (CDRs) to ascertain exposed populations and their movement (see Box 2.4) (Letouzé et al. 2015).

Another innovative approach uses CDRs to determine climate-induced migration patterns in La Guajira, Colombia. An algorithm was used to determine patterns of mobile phone use when users are at home and trace the process of their relocation to other areas after a severe drought in the region (Isaacman et al. 2018). The initiative showed a linear reduction of the population of La Guajira, totalling 10% over six months (GSMA 2018). Moreover, it demonstrated that the disruption of livelihoods (at home) and the promise of new livelihoods (in the host area) mediated this migration (Calabrese et al. 2017). The study demonstrates that CDRs are a powerful source of dynamic and granular information to understand population dynamics in the face of climate change (Isaacman et al. 2018). The initiative helps improve understanding of which parts of the population are impacted by what kinds of disturbances and the changes that result, providing insights into the nature of exposure and the sensitivity and adaptive capacity of particular groups. Crucially, data from such approaches act as a “magnifier of survey and census data, and not a substitute for them” (Letouzé et al. 2015: 13).

During the COVID-19 pandemic, several initiatives employed technology to determine the exposure of individuals to the virus in ways similar to those discussed in this section. For example, a government-mandated “contact tracing” application was downloaded on mobile phones in India; this employed Bluetooth (a wireless technology used for exchanging data) to interact with the same app on other mobile phones (The Hindu 2020). If any of the owners of the mobile phones tested positive for COVID-19, then all those who had been in contact with this individual were informed immediately and asked to quarantine.

BOX 2.4 Using mobile phones to track exposure and migration in Bangladesh

Identifying where exposed people are located is a crucial step in reducing risk and building resilience. This has been difficult historically as censuses are conducted infrequently and therefore can contain data that is outdated. Moreover, conducting robust surveys at scale can be expensive, time-consuming, arduous and, at times, impossible (e.g. in the immediate aftermath of disasters). To overcome this, a unique initiative undertaken in Bangladesh has demonstrated that CDRs (i.e. the unique, time-stamped and georeferenced record of each call made on a mobile phone) can be effective in locating exposed populations and their movements (Lu et al. 2016).

In partnership with Grameenphone, one of the largest mobile phone network providers in Bangladesh (a country that enjoys 89% mobile phone penetration) a team of researchers tracked movements of individuals in areas affected by Cyclone Mahasen in 2013 (Flowminder 2015). Using anonymised data sets from 6 million mobile phone users over a three-month period, the initiative identified high-risk behaviours such as delays in evacuation among significant numbers of exposed individuals and the sub-optimal use of cyclone shelters (Lu et al. 2016). It also challenged commonly held assumptions of mass rural–urban migration in the aftermath of extreme events, as the cyclone did not affect yearly migration patterns (Lu et al. 2016).

More importantly, the initiative demonstrates that mobile phone data can be employed to acquire granular as well as temporally and geographically precise information on the exposure of individuals to climate-induced extreme events. It also underlines how "mobile network data is a highly promising data source to supplement current survey based approaches to monitor, interpret and respond to migration from climate change, both with regard to extreme weather and slow-onset climatic stressors" (Lu et al. 2016: 6).

This section illustrated a whole range of approaches that are using data from mobile phones in innovative ways to explore how and where populations are located and the degree to which they might be exposed (see Lai et al. 2019; Laczko and Rango 2014). These approaches help overcome the problems with granularity and veracity described in Section 2.2.

2.3.3 Vulnerability

Section 2.1 outlined the way data from SRS, census or other socio-economic surveys and participatory methods can be used to better understand vulnerability. Section 2.2 highlighted that participatory methods suffer from lack of rigour in sampling and can exclude marginalised sections, that they can be manipulated by facilitators and that they are not inherently geared to operate within socially heterogenous urban contexts. Data from SRS, census and socio-economic surveys can also suffer from the lack of resolution, dynamism and reliability. Just as with hazards and exposure, a set of emergent approaches that rely on new technology are demonstrating ways to overcome these limitations.

There are a range of approaches that employ user-generated data to understand how urban residents are sensitive to climate change and to assess their capacity to adapt to its impacts. These are better able to disaggregate some of the causal factors differentiating vulnerability, such as gender, race, poverty and social networks. One initiative used data on bank card payments and ATM cash withdrawals to understand vulnerability with high spatial and temporal resolution. Examining data from 25,000 anonymised transactions from a representative sample of 100,000 individuals (representing three income groups) in the Mexican state of Baja California Sur, Martínez and colleagues (2016) learned how banking behaviour changed before, during and after the impact of Hurricane Odile in September 2014. This permitted the tracking of gender-disaggregated preparedness patterns (by, for instance, analysing individual spending on stockpiling essential goods as well as cash withdrawals for anticipated emergency expenditure). It also provided a unique insight into how much time it took various groups under analysis to recover (return to pre-disaster transaction patterns) and the relative impact of the shock (by analysing abated transactions relative to pre-disaster patterns). The study found that women increased their expenditures in preparation for the disaster twice as much as men; that those in the low-income group returned to pre-disaster transaction patterns faster than the medium- and high-income groups (probably due to the fact that their spending on stockpiling was low); and that women took much longer than men to recover.

Such georeferenced, granular data collected at scale provides resilience planners with a whole range of objective insights. For instance, it shed light on dimensions of adaptive capacity (e.g. groups ability to stockpile) and sensitivity (e.g. gauging impact through reduced transactions by population clusters relative to other clusters in the same income group; analysing the differences between income groups). This is particularly useful in urban contexts that have a significantly higher concentration of sources from which to collect big data on financial transactions (i.e. points of sale, ATMs, etc.) as compared to rural areas (Dubbudu 2017).

A growing number of innovative initiatives aim to determine the depth and dimensions of poverty using CDRs and other forms of mobile phone data, providing critical insights into climate-related vulnerability (Hardoy and Pandiella 2009). An example in east Africa has demonstrated that airtime credit purchases and patterns of mobile phone use provide valuable real-time information on several indicators related to food security which "could be integrated with early warning and monitoring systems, filling data gaps between survey intervals, and in situations where timely data is not possible or accessible" (UN Global Pulse 2015: 1). This initiative used anonymised CDRs and airtime credit purchases (time, date and value of each purchase) as proxies for food security and compared the results with those from a country-level food security survey (n = 7,500) undertaken during the same period (Decuyper et al. 2014). The results show a statistically significant correlation between airtime credit purchases and consumption of market-bought food items, thus demonstrating a pathway for determining non-monetary poverty at fine spatial scales. This means that mobile top-up data could be used as a proxy for food expenditures in market-dependent households and be used alongside other methods to gauge food security (UN Global Pulse 2015).

In another example, anonymised CDRs from two separate urban regions were used to deduce levels of poverty by analysing mobile phone use and ground truthing results with existing secondary data on poverty in these locations (Letouzé et al. 2015). The initiative used aggregate mobile phone communications activity, the level of communication between key regions and the ones under study, diversity of connections made and the level of introversion (i.e. the relative volume of calls *between* one area and another compared to the traffic *within* areas) to estimate poverty (Smith-Clarke et al. 2014) and found that it is possible to derive accurate poverty estimates using this data. This provides a cost-effective method of determining the levels and nature of poverty. Given significant levels of correlation between vulnerability and poverty, this can contribute to the development of dynamic and granular estimation of vulnerability in particular areas.

Apart from understanding vulnerability through poverty proxies, mobile phones are being used to directly measure different dimensions of vulnerability. For instance, mobile phone-based surveys were used as part of the BRACED (Building Resilience and Adaptation to Climate Extremes and Disasters) programme in Myanmar (Jones and Ballon 2020). The surveys were employed before and after shocks to acquire an in-depth understanding of the way communities are impacted by climate change and how they perceive their own "subjective" resilience and vulnerability (Jones and Tanner 2017). This has generated granular data that shed light on different aspects of vulnerability and resilience, including: ability to shift livelihoods strategies to deal with climate impacts; access to financial, social and political capital; learning capacity and preparedness and planning capacity (Jones and Tanner 2017). Access to this high-resolution and dynamic longitudinal data has immense potential to inform resilience-building actions.

In addition, machine learning is increasingly providing insights into dimensions of vulnerability. For instance, one application uses neural networks to analyse data from satellite imagery and combines this with socio-economic survey data to deliver estimates of poverty as proxies for vulnerability (Jean et al. 2016). Another approach analyses images from Google Street View using algorithms to determine income levels of different neighbourhoods (Glaeser et al. 2018). There has also been a rise in machine learning applications that use natural language processing techniques to mine data from social media content to determine vulnerability. These make it possible to use georeferenced data to "catalog 'bottom-up' perspectives on various components of vulnerability: What are the most frequently talked about climatic risks? Are certain locations consistently identified as being vulnerable in online discussions?"

(Ford et al. 2016: 10730). In relation to the COVID-19 pandemic, similar applications have been pioneered by organisations such as the World Bank, using parameters such as density of settlement and availability of community sanitation infrastructure to determine vulnerability to the virus.

2.4 CONCLUSION

This chapter made the case that emerging innovative approaches can be used to overcome the limitations in dominant approaches to data collection and analysis. It showed how AI, the dynamic use of multiple data streams and crowdsourced data can provide more reliable estimates of how and when varied climate impacts will be felt. It also illustrated how volunteer geographic information approaches and CDRs can provide dependable and granular estimates of the populations and assets that are exposed. Moreover, it exemplified the way mobile phone-administered surveys and financial transaction data can help pinpoint pockets of vulnerability with a high degree of veracity (see Table 2.1 for more).

The chapter has not argued for a wholesale rejection of existing methods, but calls for these to be complemented with data from innovative approaches. AI has the potential to enhance the value of climate models, overcoming limitations and reducing uncertainty in the model projections. CDRs can provide reliable and granular information on exposure, but this needs to be analysed alongside data from more traditional sources to enhance rigour. Similarly, while data on mobile phone use may provide reliable information on poverty, additional participatory approaches may be needed to understand the social, economic and political marginalisation that results from poverty and how this, in turn, may enhance people's sensitivity to climate impacts or influence their capacity to adapt.

We are cognisant that despite their potential, such innovative approaches come with their own limitations. First, a substantial amount of effort is needed to make sense of the large amounts of data and information acquired cheaply and swiftly through such approaches (Ford et al. 2016). For researchers mapping poverty using mobile phone data, a substantial amount of secondary research and ground truthing was needed to decipher levels of poverty (Smith-Clarke et al. 2014). In the absence of these additional layers of analysis, these methods can deliver an incomplete picture of exposure, vulnerability and climate hazards.

TABLE 2.1 Innovative data approaches to overcome limitations of orthodox methods

	Certainty	Granularity	Veracity
Hazards	• AI can help integrate simulations of complicated land–air–water interactions at fine spatial scales into global climate models, enhancing the certainty of the projections that they deliver.	• Dynamic use of large amounts of data from multiple streams provides precision forecasts on extreme events.	• Crowdsourced data on urban microclimates using algorithms built through rigorous calibration can provide reliable data on heterogenous urban environments.
Exposure	• Developing probabilistic hazard models using information from open mapping approaches provides clarity and precision on populations exposed to different hazards.	• Volunteer geographic information approaches can help provide dynamic real-time information on exposed populations and assets at extremely fine spatial scales.	• Analysis of CDRs can provide reliable real-time information on exposed populations cheaply and swiftly.
Vulnerability	• Remotely administered surveys can provide longitudinal real-time information on populations that are sensitive to different climate impacts and their capacity to adapt to these.	• Analysis of large volumes data on financial transactions have helped pinpoint populations that are more vulnerable than others.	• Analysing data on mobile phone usage and airtime credit purchases provides reliable information on vulnerability.

Second, there is a growing concern over the equity dimensions of using big data approaches and applications. Letouzé et al. (2015) explored the different dimensions of this and found that low-income countries have only 1% of the world's capacity to transmit data via high-speed internet. Along with this, certain applications discussed in Section 2.3 (e.g. preparation of maps using drones) requires expertise that may not exist in the context of low-income countries. Therefore, some of the world's least resilient contexts would derive relatively less benefit if such approaches proliferate without a proportionate increase in internet capacities and expertise.

Third, despite the large sample sizes in big data, sampling biases can exist, as age, income and geographic location can all determine the degree to which people use mobile phones and other forms of technology that generate data. This highlights that use of technology does not reduce the importance of robust sampling strategies.

Fourth, over-reliance on these approaches poses the risk that decision makers will overlook local engagement and tailored strategies for acquiring data and information. This is a particular problem for data on vulnerability and exposure as these have strong sociopolitical drivers that have proven effective in understanding why and how people are impacted by climate change.

Finally, it is also important to acknowledge that the acquisition of high-quality and decision-friendly information and data is not a guarantee of its effective use in policy formulation. This requires navigating through institutional politics, building contextually relevant narratives of change, mobilising coalitions of advocates and finding leaders to act as policy entrepreneurs for effecting change (Bahadur and Tanner 2014; Tanner et al. 2019).

In conclusion, it is important to understand that the arguments presented in this chapter are particularly salient for urban contexts, as there are more opportunities in towns and cities than in rural contexts for the use of innovative approaches. Most innovative approaches rely on the acquisition and analysis of large amounts of data and information. This is generated from nodes such as mobile phones, computers, point-of-sales machines, ATMs, etc., that are concentrated in urban areas. For instance, in India, compared to rural citizens 22% more urban dwellers own mobile phones (ET Bureau 2018). Similarly, India's urban centres have almost 40% more ATMs than its villages (Dubbudu 2017). Also, certain innovative data are reliant on social media platforms, and content on these platforms is predominantly generated by urban users. This apart, with most of the world's universities, research institutes and trained professionals, urban centres are sites of innovation and have the ability to deploy innovative approaches for understanding hazards, exposure and vulnerability better (Leichenko 2011).

REFERENCES

Adiningsih, E. (2010, 9–18 June). *The applications of satellite remote sensing on climate change and food security in Indonesia* [Paper presentation]. 53rd Session of UN-COPUOS. Vienna.

African Development Bank. (2014). *Tracking Africa's Progress in Figures.* Tunis: African Development Bank.

Bader, D. A., Blake, R., Grimm, A., Hamdi, R., Kim, Y., Horton, R., Rosenzweig, C., Alverson, K., Gaffin, S. and Crane, S. (2018). Urban climate science. In C. Rosenzweig, W. Solecki, P. Romero-Lankao, S. Mehrotra, S. Dhakal and S. Ali Ibrahim (Eds), *Climate Change and Cities: Second Assessment Report of the Urban Climate Change Research Network* (pp. 27–60). New York: Urban Climate Change Research Network.

Bader, D. C., Covey, C., Gutowski, W. J., Jr, Held, I. M., Miller, R. L., Tokmakian, R. T. and Zhang, M. H. (2008). *Climate Models: An Assessment of Strengths and Limitations.* Washington DC: US Department of Energy Publications.

Bahadur, A. V. (2014). *Policy climates and climate policies: Analysing the politics of building resilience to climate change* [PhD thesis]. University of Sussex, UK.

Bahadur, A. V. and Tanner, T. (2014). Policy climates and climate policies: Analysing the politics of building urban climate change resilience. *Urban Climate*, *7*, 20–32.

Bahadur, A., Tanner, T. and Pichon, F. (2016). *Enhancing Urban Climate Change Resilience: Seven Entry Points for Action.* Manila: Asian Development Bank.

Brown, D., Howes, M., Hussein, K., Longley, C. and Swindell, K. (2002). *Participatory Methodologies and Participatory Practices: Assessing PRA Use in the Gambia.* Agricultural Research and Extension Network Paper 124. London: Overseas Development Institute.

Bushwick, S. (2013). *How to monitor urban weather with your smartphone.* Inside Science. Retrieved 23 November, 2020, from https://www.insidescience.org/news/how-monitor-urban-weatherwith-your-smartphone

Cadag, J. R. D. and Gaillard, J. C. (2011). Integrating knowledge and actions in disaster risk reduction: The contribution of participatory mapping. *Area*, *44*(1), 100–109.

Calabrese, F., Moro, E., Blondel, V. and Pentland, A. (Eds). (2017). *Netmob Book of Abstracts.* Milan: Netmob.

Cardona, O. D., van Aalst, M. K., Birkmann, M., Fordham, M., McGregor, G., Perez, R., Pulwarty, R. S., Schipper, E. L. F. and Sinh, B. T. (2012). Determinants of risk: Exposure and vulnerability. In C. B. Field, V. Barros, T. F. Stocker, D. Qin, D. J. Dokken,

K. L. Ebi, M. D. Mastrandrea, K. J. Mach, G.-K. Plattner, S. K. Allen, M. Tignor and P. M. Midgley (Eds), *Managing the Risks of Extreme Events and Disasters to Advance Climate Change Adaptation: Special Report of Working Groups I and II of the Intergovernmental Panel on Climate Change* (pp. 65–108). Cambridge, UK: Cambridge University Press.

Center For International Earth Science Information Network. (2013). *Low Elevation Coastal Zone (LECZ) Urban-Rural Population and Land Area Estimates, Version 2*. Palisades, NY: NASA Socioeconomic Data and Applications Center (SEDAC). Retrieved 23 November, 2020, from https://sedac.ciesin.columbia.edu/data/set/lecz-urban-rural-population-land-area-estimates-v2

Ceola, S., Laio, F. and Montanari, A. (2014). Satellite nighttime lights reveal increasing human exposure to floods worldwide. *Geophysical Research Letters, 41*(20), 7184–7190.

Chiwaka, E. and Yates, R. (2012). *Participatory Vulnerability Analysis*. Johannesburg: ActionAid International.

Connelly, A., Carter, J., Handley, J. and Hincks, S. (2018). Enhancing the practical utility of risk assessments in climate change adaptation. *Sustainability 10*(5), Article 1399. https://doi.org/10.3390/su10051399

Cox, J. R., Rosenzweig, C., Solecki, W., Goldberg, R. and Kinney, P. (2006). *Social Vulnerability to Climate Change: A Neighborhood Analysis of the Northeast US Megaregion*. Draft white paper for Union of Concerned Scientists, Northeast Climate Change Impact Study.

De Blasio, F. V. (2011). Landslides in Valles Marineris (Mars): A possible role of basal lubrication by sub-surface ice. *Planetary and Space Science, 59*(13), 1384–1392.

Decuyper, A., Rutherford, A., Wadhwa, A., Bauer, J. M., Krings, G., Gutierrez, T., Blondel, V. D. and Luengo-Oroz, M. A. (2014). Estimating food consumption and poverty indices with mobile phone data. *arXiv 2014*, 1412.2595.

Deville, P., Linard, C., Martin, S., Gilbert, M., Stevens, F. R., Gaughan, A. E., Blondel, V. D. and Tatem, A. J. (2014). Dynamic population mapping using mobile phone data. *Proceedings of the National Academy of Sciences USA 111*(45), 15888–15893.

Dickson, E., Baker, J. L., Hoornweg, D. and Asmita, T. (2012). *Urban Risk Assessments: An Approach for Understanding Disaster and Climate Risk in Cities*. Washington DC: The World Bank.

Djoudi, H., Locatelli, B., Vaast, C., Asher, K., Brockhaus, M. and Sijapati, B. B. (2016). Beyond dichotomies: Gender and intersecting inequalities in climate change studies. *Ambio, 45*(3), 248–262.

Dubbudu, R. (2017). *Number of ATMs in Rural areas less than what it was in September 2016*. Factly. Retrieved 23 November, 2020, from https://factly.in/number-atms-rural-areas-less-september-2016/

Eastin, J. (2018). Climate change and gender equality in developing states. *World Development*, *107*, 289–305.

Ebert, A., Kerle, N. and Stein, A. (2009). Urban social vulnerability assessment with physical proxies and spatial metrics derived from air- and spaceborne imagery and GIS data. *Natural Hazards*, *48*, 275–294.

ET Bureau. (2018, 7 August). *Urban-rural mobile ownership gap: India below Pak, B'desh*. The Economic Times. Retrieved 23 November, 2020, from https://economictimes.indiatimes.com/tech/hardware/urban-rural-mobile-ownership-gap-india-below-pak-bdesh/articleshow/65313802.cms

Flower, B., Fortnam, M., Kol, L., Sasin, P. and Wood, R. G. 2018. Using participatory methods to uncover interacting urban risks: A case study of three informal settlements in Phnom Penh, Cambodia. *Environment and Urbanization*, *30*(1), 301–316.

Flowminder. (2015). *Mobile phone data to understand climate change and migration patterns in Bangladesh*. Retrieved 23 November, 2020, from https://web.flowminder.org/case-studies/mobile-phone-data-to-understand-climate-change-and-migration-patterns-in-bangladesh

Ford, J. D., Tilleard, S. E., Berrang-Ford, L., Araos, M., Biesbroek, R., Lesnikowski, A. C., MacDonald, G. K., Hsu, A., Chen, C. and Bizikova, L. (2016). Opinion: Big data has big potential for applications to climate change adaptation. *Proceedings of the National Academy of Sciences USA*, *113*(39), 10729–10732.

Gencer, E., Folorunsho, R., Linkin, M., Wang, X., Natenzon, C. E., Wajih, S., Mani, M., Esquivel, M., Ali Ibrahim, S., Tsuneki, H., Castro, R., Leone, M. F., Dilnoor, P., Romero-Lankao, P., Solecki, W., Lin, B. and Panda, A. (2018). Disasters and risk in cities. In C. Rosenzweig, W. D. Solecki, P. Romero-Lankao, S. Mehrotra, S. Dhakal and S. Ali Ibrahim (Eds), *Climate Change and Cities* (pp. 61–98). New York: Cambridge University Press.

Gentine, P., Pritchard, M., Rasp, S., Reinaudi, G. and Yacalis, G. (2018). Could machine learning break the convection parameterization deadlock? *Geophysical Research Letters*, *45*(11), 5742–5751.

GIZ. (2014). *A Framework for Climate Change Vulnerability Assessments*. Bonn: Deutsche Gesellschaft für Internationale Zusammenarbeit (GIZ).

Glaeser, E. L., Kominers, S. D., Luca, M. and Naik, N. (2018). Big data and big cities: The promises and limitations of improved measures of urban life. *Economic Inquiry*, *56*(1), 114–137.

Government of India. (2011). *Houselisting and Housing Census Data – 2011?* Census of India. Retrieved 23 November, 2020, from http://censusindia.gov.in/2011census/hlo/HLO_Tables.html

Groth, M., Brück, M. and Oberascher, T. (2016). Climate change related risks, opportunities and adaptation actions in European cities – Insights from responses to the CDP cities program. ERSA conference papers ersa16p17, European Regional Science Association. Retrieved 23 November, 2020, from www-sre.wu.ac.at/ersa/ersaconfs/ersa16/Paper17_MarkusGroth.pdf

GSMA. (2018). *Big data for social good.* Retrieved 23 November, 2020, from https://www.gsma.com/newsroom/blog/big-data-social-good-achievements-one-year-looking-ahead-mobile-world-congress–2018/

Guzzetti, F., Mondini, A. C., Cardinali, M., Fiorucci, F., Santangelo, M. and Chang, K. T. (2012). Landslide inventory maps: New tools for an old problem. *Earth-Science Reviews*, *112*(1–2), 42–66.

Hardoy, J. and Pandiella, G. (2009). Urban poverty and vulnerability to climate change in Latin America. *Environment and Urbanization*, *21*(1), 203–224.

Harrison, P. A., Holman, I. P. and Berry, P. M. (2015). Assessing cross-sectoral climate change impacts, vulnerability and adaptation: An introduction to the CLIMSAVE project. *Climatic Change*, *128*, 153–167.

Harvey, M., Eltinay, N., Barnes, S., Guerriero, R. and Caffa, M. (2018). *Open Data Infrastructure for City Resilience: A Roadmap Showcase and Guide*. Geneva: United Nations.

Herron, H., Bohn, B., Roy, S. and Evans, W. (2014). *Climate Change Data and Risk Assessment Methodologies for the Caribbean.* Washington DC: Inter-American Development Bank.

Hickey, S. and Mohan, G. (2004). *Participation – from Tyranny to Transformation? Exploring New Approaches to Participation in Development*. London: Zed Books.

IFRC. (1999). *Vulnerability and Capacity Analysis*. Geneva: International Federation of Red Cross and Red Crescent Societies.

Intergovernmental Panel on Climate Change. (2014). Summary for policymakers. In C. B. Field, V. R. Barros, D. J. Dokken, K. J. Mach, M. D. Mastrandrea, T. E. Bilir, M. Chatterjee, K. L. Ebi, Y. O. Estrada, R. C. Genova, B. Girma, E. S. Kissel, A. N. Levy, S. MacCracken, P. R. Mastrandrea and L. L. White (Eds), *Climate Change 2014: Impacts, Adaptation, and Vulnerability. Part A: Global and Sectoral Aspects. Contribution of Working Group II to the Fifth Assessment Report of the Intergovernmental Panel on Climate Change* (pp. 1–32). New York: Cambridge University Press.

Intergovernmental Panel on Climate Change. (2019). *What is a GCM?* Retrieved 23 November, 2020, from https://www.ipcc-data.org/guidelines/pages/gcm_guide.html

Isaacman, S., Frias-Martinez, V. and Frias-Martinez, E. (2018). Modelling human migration patterns during drought conditions in La Guajira, Colombia. In *Proceedings of the 1st ACM SIGCAS Conference on Computing and Sustainable Societies (COMPASS) 2018*, Article 31. New York: Association for Computing Machinery. https://doi.org/10.1145/3209811.3209861

Jean, N., Burke, M., Xie, M., Davis, W. M., Lobell, D. B. and Ermon, S. (2016). Combining satellite imagery and machine learning to predict poverty. *Science*, *353*(6301), 790–794.

Jones, L. (2018, 10 December). Can artificial intelligence help build better, smarter climate models? *Yale Environment 360*. Retrieved 23 November 2020, from https://e360.yale.edu/features/can-artificial-intelligence-help-build-better-smarter-climate-models

Jones, L. and Ballon, P. (2020). Tracking changes in resilience and recovery after natural hazards: Insights from a high-frequency mobile-phone panel survey. *Global Environmental Change*, *62*, Article 102053.

Jones, L., Dougill, A., Jones, R. G., Steynor, A., Watkiss, P., Kane, C., Koelle, B., Moufouma-Okia, W., Padgham, J., Ranger, N., Roux, J.-P., Suarez, P., Tanner, T. and Vincent, K. (2015). Ensuring climate information guides long-term development. *Nature Climate Change*, *5*, 812–814.

Jones, L. and Tanner, T. (2017). "Subjective resilience": Using perceptions to quantify household resilience to climate extremes and disasters. *Regional Environmental Change*, *17*(1), 229–243.

Kaku, K. (2019). Satellite remote sensing for disaster management support: A holistic and staged approach based on case studies in Sentinel Asia. *International Journal of Disaster Risk Reduction*, *33*, 417–432.

Knippenberg, E., Jensen, N. and Constas, M. (2019). Quantifying household resilience with high frequency data: Temporal dynamics and methodological options. *World Development*, *121*, 1–15.

Korf, B. (2002). *Does PRA Make Sense in Democratic Societies?* London: International Institute of Environment and Development.

Laczko, F. and Rango, M. (2014). Can big data help us achieve a "migration data revolution"? *Migration Policy Practice (IOM)*, *4*(2), 20–29.

Lai, S., zu Erbach-Schoenberg, E., Pezzulo, C., Ruktanonchai, N. W., Sorichetta, A., Steele, J., Li, T., Dooley, C. A. and Tatem, A. J. (2019). Exploring the use of mobile phone data for national migration statistics. *Palgrave Communications*, *5*, Article 34.

Lankao, P.R. and Qin, H. (2011). Conceptualizing urban vulnerability to global climate and environmental change. *Current opinion in environmental sustainability*, *3*(3), 142–149.

Lee, T. and Hughes, S. (2017). Perceptions of urban climate hazards and their effects on adaptation agendas. *Mitigation and Adaptation Strategies for Global Change*, *22*(5), 761–776.

Leichenko, R. (2011). Climate change and urban resilience. *Current Opinion in Environmental Sustainability*, *3*(3), 164–168.

Letouzé, E., Vinck, P., Schwarz, B., Sala, S., Sangkoyo, D. and Tellman, T. (2015). *Big Data for Climate Resilience*. London: Harvard Humanitarian Initiative, MIT Media Lab and Overseas Development Institute.

Leurs, R. (1996). Current challenges facing participatory rural appraisal. *Public Administration and Development*, *16*(1), 57–72.

Lu, X., Wrathall, D. J., Sundsøy, P. R., Nadiruzzaman, M., Wetter, E., Iqbal, A., Qureshi, T., Tatem, A., Canright, G., Engø-Monsen, K. and Bengtsson, L. (2016). Unveiling hidden migration and mobility patterns in climate stressed regions: A longitudinal study of six million anonymous mobile phone users in Bangladesh. *Global Environmental Change*, *38*, 1–7. https://doi.org/10.1016/j.gloenvcha.2016.02.002

Lund, J. F. and Saito-Jensen, M. (2013). Revisiting the issue of elite capture of participatory initiatives. *World Development*, *46*, 104–112. https://doi.org/10.1016/j.worlddev.2013.01.028

Lynch, P. (2010). Weather and climate forecasting: Chronicle of a revolution. *WMO Bulletin*, *59*(2), 75–78.

Martínez, E. A., Rubio, M. H., Martinez, R. M., Arias, J. M., Patane, D., Zerbe, A., Kirkpatrick, R. and Luengo-Oroz, M. (2016, 25 September). *Measuring economic resilience to natural disasters with big economic transaction data* [Paper presentation]. Bloomberg Data for Good Exchange Conference, New York.

Maslin, M. (2013). Cascading uncertainty in climate change models and its implications for policy. *The Geographical Journal*, *179*(3), 264–271.

McCarthy, J. J., Canziani, O. F., Leary, N. A., Dokken, D. J. and White, K. S. (Eds). (2001). *Climate Change 2001: Impacts, Adaptation, and Vulnerability: Contribution of Working Group II to the Third Assessment Report of the Intergovernmental Panel on Climate Change* (Vol. 2). Cambridge: Cambridge University Press.

Mehrotra, S., Natenzon, C. E., Omojola, A., Folorunsho, R., Gilbride, J. and Rosenzweig, C. (2009). *Framework for city climate risk assessment*. Fifth Urban Research Symposium. Marseille: World Bank.

Moser, C. and Stein, A. (2011). Implementing urban participatory climate change adaptation appraisals: A methodological guideline. *Environment and Urbanization*, *23*(2), 463–485.

Natty, M. (2018). *Ramani Huria – Mapping Innovations and Community Data In Dar Es Salaam*. Global Facility for Disaster Risk Reduction. Retrieved 23 November, 2020, from https://www.gfdrr.org/sites/default/files/Ramani%20Huria%20-%20Mapping%20Innovations%20and%20Community%20Data%20in%20Dar%20Es%20Salaam.pdf

Naylor, L. A., Spencer, T., Lane, S. N., Darby, S. E., Magilligan, F. J., Macklin, M. G. and Möller, I. (2017). Stormy geomorphology: Geomorphic contributions in an age of climate extremes. *Earth Surface Processes and Landforms*, *42*(1), 166–190.

Ospina, A. (2018). Emerging stories of big data for resilience building: Dar Ramani Huria. *International Institute for Sustainable Development*. Retrieved 23 November, 2020, from https://www.iisd.org/blog/emerging-stories-big-data-resilience-building-dar-ramani-huria#_edn1

Overeem, A., Droste, A. M., Pape, J. J., Leijnse, H., Steeneveld, G. J., Van Delden, A. J. and Uijlenhoet, R. (2017). Crowdsourcing urban air temperatures through smartphone battery temperatures in São Paulo, Brazil. *Journal of Atmospheric and Oceanic Technology*, *34*(9), 1853–1866.

Overeem, A., Robinson, J. C. R., Leijnse, H., Steeneveld, G. J., Horn, B. K.P. and Uijlenhoet, R. (2013). Crowdsourcing urban air temperatures from smartphone battery temperatures. *Geophysical Research Letters*, *40*(15), 4081–4085.

Piccolella, A. (2013). Participatory mapping for adaptation to climate change: The case of Boe Boe, Solomon Islands. *Knowledge Management for Development Journal*, *9*(1), 24–36.

Raghupathi, W. and Raghupathi, V. (2014). Big data analytics in healthcare: Promise and potential. *Health Information Science and Systems*, *2*, Article 3.

Randall, D. A., Wood, R. A., Bony, S., Colman, R., Fichefet, T., Fyfe, J., Kattsov, V., Pitman, A., Shukla, J., Srinivasan, J. and Stouffer, R. J. (2007). Climate models and their evaluation. In S. Solomon, D. Qin, M. Manning, M. Marquis, K. Averyt, M. Tignor, H. Miller and Z. Chen (Eds), *Climate Change 2007: The Physical Science Basis. Contribution of Working Group I to the Fourth Assessment Report of the IPCC (FAR)* (pp. 589–662). Cambridge: Cambridge University Press.

Reliefweb. (2011). *Tanzania: Floods and landslides*. Retrieved 23 November, 2020, from https://reliefweb.int/disaster/fl-2011-000183-tza

Reliefweb. (2014). *Tanzania: Floods*. Retrieved 23 November, 2020, from https://reliefweb.int/disaster/fl–2014–000053-tza

Revi, A., Satterthwaite, D. E., Aragón-Durand, F., Corfee-Morlot, J., Kiunsi, R.B.R., Pelling, M., Roberts, D.C. and Solecki, W. (2014). Urban areas. In C. B. Field, V. R. Barros, D. J. Dokken, K. J. Mach, M. D. Mastrandrea, T. E. Bilir, M. Chatterjee, K. L. Ebi, Y. O. Estrada, R. C. Genova, B. Girma, E. S. Kissel, A. N. Levy, S. MacCracken, P. R. Mastrandrea and L. L. White (Eds), *Climate Change 2014: Impacts, Adaptation, and Vulnerability. Part A: Global and Sectoral Aspects. Contribution of Working Group II to the Fifth Assessment Report of the Intergovernmental Panel on Climate Change* (pp. 535–612). Cambridge: Cambridge University Press.

Ruf, C., Unwin, M., Dickinson, J., Rose, R., Rose, D., Vincent, M. and Lyons, A. (2013). CYGNSS: Enabling the future of hurricane prediction [remote sensing satellites]. *IEEE Geoscience and Remote Sensing Magazine*, *1*(2), 52–67.

Scoones, I. (2019). *What Is Uncertainty and Why Does It Matter?* Brighton: Institute of Development Studies.

Smith-Clarke, C., Mashhadi, A. and Capra, L. (2014). Poverty on the cheap: Estimating poverty maps using aggregated mobile communication networks. In *CHI'14: Proceedings of the SIGCHI Conference on Human Factors in Computing Systems* (pp. 511–520). New York: Association for Computing Machinery. https://doi.org/10.1145/2556288.2557358

Soesilo, D. (2015). *Obtaining High-Resolution Imagery to Map and Model Flood Risks in Dar es Salaam*. Geneva: Swiss Foundation for Mine Action.

Tanner, T., Acharya, S. and Bahadur, A. (2018). *Perceptions of Climate Change: Applying Assessments to Policy and Practice*. Delhi: Oxford Policy Management and Practical Action.

Tanner, T., Zaman, R. U., Acharya, S., Gogoi, E. and Bahadur, A. (2019). Influencing resilience: The role of policy entrepreneurs in mainstreaming climate adaptation. *Disasters*, *43*(S3), S388–S411.

The Hindu. (2020). *How does the Aarogya Setu app work?* Retrieved 23 November, 2020, from https://www.thehindu.com/news/national/how-does-the-aarogya-setu-app-work/article31532073.ece

Tooth, S. and Viles, H. (2014). *10 Reasons Why Geomorphology Is Important*. London: British Society for Geomorphology.

Tralli, D. M., Blom, R. G., Zlotnicki, V., Donnellan, A. and Evans, D. L. (2005). Satellite remote sensing of earthquake, volcano, flood, landslide and coastal inundation hazards. *ISPRS Journal of Photogrammetry and Remote Sensing*, *59*(4), 185–198.

Turnbull, M. and Turvill, E. (2012). *Participatory Capacity and Vulnerability Analysis: A Practitioner's Guide*. Oxford: Oxfam.

UN Global Pulse. (2015). *Using Mobile Phone Data and Airtime Credit Purchases to Estimate Food Security*. New York: UN Global Pulse Project.

United States Army Corps of Engineers. (2006). *Depth-Damage Relationships for Structures, Contents, and Vehicles and Content-to-Structure Value Ratios (CSVR) in Support of the Donaldsonville to the Gulf, Louisiana, Feasibility Study*. New Orleans: United States Army Corps of Engineers.

United States Geological Survey. (2019) *What is a topographic map?* Retrieved 23 November, 2020, from https://www.usgs.gov/faqs/what-a-topographic-map

Vallgren, A. and Petrykowska, L. (2016). *Ignitia Tropical Weather Forecasting*. Accra: Ignitia Ghana Ltd.

van Oldenborgh, G. J., Reyes, F. D., Drijfhout, S. S. and Hawkins, E. (2013). Reliability of regional climate model trends. *Environmental Research Letters*, *8*(1), Article 014055.

Wang, Y., Zhao, L., Yang, D. and Moses, M. (2015). GIS-based climate change vulnerability mapping at the urban scale: A case study of Shanghai metropolitan area in China. *International Journal of Environmental Studies*, *72*(6), 1002–1016.

Ward, J. (2018). *Will future megacities be a marvel or a mess?* Bloomberg. Retrieved 23 November, 2020, from https://www.bloomberg.com/news/features/2018-11-02/india-s-new-delhi-is-example-how-urbanization-leads-to-megacities

Willows, R. (2003). *Climate Adaptation Risk, Uncertainty and Decision-Making*. Oxford: UK Climate Impacts Programme.

World Bank. (2016). *The Atlas of Flood Resilience in Dar es Salaam*. Washington DC: World Bank Group.

World Meteorological Association. (2019). *Commission for Climatology*. Retrieved 23 November, 2020, from www.wmo.int/pages/prog/wcp/ccl/faqs.php

Chapter 3

Resilient urban communities: From incremental to transformational change

THE field of urban resilience has evolved over the past two decades. As discussed in Chapter 1, the first phase of urban resilience research and practice focused primarily on terrorism and security concerns. Resilience was considered a particular type of response to specific shocks rather than an approach to managing systems so as to allow them to function despite a variety of shocks and stresses. In the second phase, urban resilience starts to actively consider impacts emanating from climate change using primary empirical data. In the third phase, resilience is firmly recognised as an approach to inform the governance of cities.

The concept of urban resilience matures in this third phase and practical initiatives emerge, especially those aiming to enhance the ability of vulnerable communities to deal with climate-induced shocks and stresses. This draws heavily on pre-existing approaches of community-based disaster risk reduction and community-based adaptation, defined as "a form of adaptation that aims to reduce the risks of climate change to the world's poorest people by involving them in the practices and planning of adaptation" (Forsyth 2013: 439). Typically, community-based adaptation projects are located in vulnerable rural areas and based on local priorities, needs, knowledge and capacities (Bharwani 2011). These entail a spectrum of actions that range from livelihood diversification and ecosystem conservation to the development of community-level protective infrastructure and capacity building (Archer et al. 2014). Urban resilience initiatives borrowed many tools and approaches from community-based adaptation to work with vulnerable communities in urban contexts.

Urban climate change resilience initiatives have emerged during a time when the scale of vulnerability and exposure to climate-induced shocks and stresses has increased substantially, a trend that is likely to continue. Currently, 55% of the world's population is concentrated in cities, and this number is likely to swell to 68% by 2050 (UN DESA 2018). Urban populations are highly vulnerable to the impacts of climate change as one in every three urban residents lives in informal settlements that are characterised by poor health indicators, lack of access to basic services and high rates of crime and violence (World Bank 2019). Moreover, 746 million urban residents live on less than USD 2 a day, with most of these concentrated in low- and middle-income countries (Garland 2007). For example, 25% of India's urban population lives below the national poverty line (Sharma 2017). Cities are also highly exposed to climate impacts, and 66% of all cities over 5 million people are located in highly exposed low-elevation coastal zones (Greenfieldboyce 2007). Almost 480 million people live in cities that are highly exposed to cyclone risk, and this figure is set to rise to 680 million by 2050 (Brecht et al. 2013). Also, in the three decades between 2000 and 2030, the amount of urban area exposed to flood and drought will increase by 250% (Güneralp et al. 2015).

This chapter argues for a significant shift in approaches to enhancing the resilience of urban communities from those that seek incremental improvement in resilience and risk management to those that demonstrate more transformational changes in the structures driving risk and constraining agency. In the following sections, we characterise the current state of play in community resilience initiatives, examine challenges to these approaches and present a pivot needed to transform urban climate change resilience going forward.

3.1 DOMINANT APPROACHES FOR ENHANCING THE RESILIENCE OF URBAN COMMUNITIES

The current constellation of actions and initiatives to enhance community resilience to the impacts of a changing climate

TABLE 3.1 Examples of existing approaches for urban community resilience

	Hazards	Exposure	Vulnerability
Building assets	Peri-urban agriculture, Sri Lanka	Storm-resilient housing, Vietnam	Microfinance for ecosystem-based adaptations, Peru and Colombia
Knowledge and awareness		Community-based early warning system, Indonesia	Water demand management, Spain
Resilient infrastructure	Lake restoration, India		Community-based water harvesting, Indonesia
Governance	• Comprehensive resilience planning by the Surat Climate Change Trust, India • Co-management of common property resources, India		

is as vast as it is varied. In general, they operate at the local level within communities that are vulnerable to climate impacts and implement community-based development initiatives that aim to enhance the adaptive capacity of local populations (Dodman and Mitlin 2013). In doing so, they draw local knowledge into decision-making and build on local cultural norms. In unpacking existing urban climate change resilience approaches at community level, this section builds on a range of reviews of community-based adaptation and disaster management to distinguish four categories of actions: building assets; knowledge and awareness; governance; and infrastructure (Ayers and Forsyth 2009; Archer et al. 2014; Paterson and Charles 2019). For each of these areas, we provide examples to paint a picture of current practice (summarised in Table 3.1).

3.1.1 Building assets for community resilience

Many urban climate change resilience interventions respond to a core imperative of securing basic material necessities. These may include food, shelter and belongings as well as livelihoods and health (Paterson and Charles 2019). Several community-based initiatives aiming to reduce vulnerability illustrate this category. One such initiative is a typhoon-resilient housing programme in Da Nang, Vietnam. A city of over 1.23 million, Da Nang is located in the South Central Coast region, which is exposed to tropical storms that have historically affected 80–90% of the population (Tran et al. 2014). In 2006, Typhoon Xangsane resulted in the complete collapse of almost 1,300 houses and damaged another 13,000; and in 2013, Typhoon Nari caused damage amounting to almost USD 40 million (Phong 2013). This apart, climate change is set to exacerbate the problem as, according to projections, the intensity of severe rainfall events (those with a return period of 50 years or more) is likely to increase by 2030 (Opitz-Stapleton 2013).

As a consequence, the Women's Union in Da Nang, in collaboration with international partners, administers a fund that makes revolving loans for enhancing the resilience of the homes of the urban poor in eight disaster-prone wards by reducing their exposure to storms (United Nations Climate Change 2020b). This initiative began with a vulnerability assessment, surveys to collect historical information about past damage and participatory meetings with marginalised communities to discuss their requirements for more resilient housing (Tran et al. 2014). This fed into the organisation of a "design competition" to find cost-effective solutions for resilient housing that saw wide participation from architectural firms and led to the selection of a resilient housing design (Tran et al. 2014). In total, 244 houses were built using revolving loans from the Women's Union. These employed the selected design, which included new approaches for reinforcing roofs and the incorporation of new kinds of posts and beams into the structure (Tran et al. 2014). The success of the initiative was shown by the fact none of the houses suffered any damage due to Typhoon Nari in 2013. In this way, the initiative drew on local knowledge and pooled community resources to reduce the vulnerability of marginalised urban communities.

Another example of an approach to improve individual and material well-being through risk reduction, enhanced food security and increasing household assets is found in Kesbewa, Sri Lanka. Kesbewa is a rapidly urbanising satellite town, on the western periphery of Colombo District, with increasing problems of saltwater intrusion (Dubbeling 2014). Moreover, increasing rainfall and the conversion of agricultural land into residential neighbourhoods has led to increased flooding. Mohamed and Gunasekera (2014) document that the provincial ministry of agriculture recognised that well-maintained paddy fields act as flood buffer zones by storing water and regulating drainage. The ministry therefore encouraged a constellation of local actors to execute pilot projects on urban agriculture to reduce the rate at which paddy fields were being converted into residential areas and to help revitalise unused paddy fields. One of the projects promoted salt-resistant varieties of rice alongside cultivation of vegetables in raised beds and involved 47 farmers in four different locations totalling 43 acres. Research on the impact of this intervention revealed that participating households were able to increase income, reduce expenditure on food, improve food security and enhance dietary diversity. Crucially, flooding incidences and impacts were lessened due to the storm water infiltration and management function of paddy fields. Pivotal to the success of this initiative was the fact that "the participation of project stakeholders has been high, with governments, agricultural institutions and the urban council taking a leading role" (Mohamed and Gunasekera 2014: 22).

Crucially, these examples do not relate only to the strengthening of material assets; there is also growing interest in the role of social capital and relational approaches. These involve collective action and seek to understand and build on interactions among community members to enhance assets for community resilience (see Box 3.1) (Aldrich and Meyer 2015; Paterson and Charles 2019).

3.1.2 Knowledge and awareness for resilience

Interventions in this category include tracking, monitoring and gauging risks to enhance awareness amongst vulnerable populations (Paterson and Charles 2019). This can be done through the use of scientific or traditional knowledge.

Community-based early warning systems are a typical example of activities in this category, and these have been implemented across the world in different forms. The Flood Early Warning Early Action System developed by a group of organisations led by the Indonesian Red Cross in Java provides a useful illustration. Java is one of Indonesia's most populous and rapidly urbanising regions and has historically witnessed annual flooding (Firman 2017). The established patterns of inundation are now shifting and "due to changing climate and environmental conditions, floods have become more extreme, with huge implications for the people who live along the rivers" (IFRC 2018: 1). To remedy this problem, the team running the project has developed an internet-based application to predict and monitor rainfall and flooding by collecting real-time rainfall, water level and weather data (Susandi and Tamamdin

BOX 3.1 Relational approaches for building assets for community resilience: Microfinance for ecosystem-based adaptation

Microfinance typically involves the generation of finance through aggregating savings from communities and then providing small, short-term loans to those in need. This kind of financing helps the poor "build up their assets, establish or further develop a business, increase their wealth, and protect against risks" (Agrawala and Cararo 2010: 14). The potential of microfinance to help vulnerable communities ameliorate climate impacts has been well recognised, because they support asset development among vulnerable households that, in turn, helps build adaptive capacity (Hammill et al. 2008). Also, microfinance provides a safety net in times of crisis, and conditionalities on loans given can encourage adaptative behaviours (CIF 2018). Recognising the potential of this instrument, the United Nations Environment Programme (UNEP) collaborated with a range of international institutions and national microfinance institutions to launch the Microfinance for Ecosystem-Based Adaptation to Climate Change project among smallholders in peri-urban areas in Peru and Colombia. As Krummheur et al. (2015) catalogue, the project stipulated that in order to access microcredit, smallholders must demonstrate how the sums borrowed will strengthen household economies, increase resilience to climate change, reduce pressure on ecosystems, reduce risks associated with climatic events in productive activities and protect or improve biodiversity. Overall, 1,200 disbursements have been made under the project in support of 40 types of ecosystem-based adaptation (e.g. drip irrigation, shift to organic agriculture, bio-digestors and rainwater harvesters). In this way, technical assistance from development actors combined with community knowledge and pooled resources are effectively being employed to support adaptation to climate change.

2017). It uses this information to deliver a green alert, yellow alert or red alert and locates risky areas using Google Maps and other GIS maps.

Crucially, teams of local Red Cross volunteers from the flood-affected communities, local government employees and disaster response agencies have been trained in using this system to undertake different sets of activities (IFRC 2018). For instance, a green alert aims to increase levels of vigilance and for moving valuables to a safer place; a yellow alert means volunteers and others part of the disaster risk management apparatus ensure that communities at risk move to higher floors of their houses or buildings; and a red alert spurs this group to help evacuate people living around the dangerous areas (Susandi and Tamamdin 2017). Combining a technology solution with community-based action in this way has led to a reduction in the risk from floods in the region, and similar systems are in place across the world, from the Caribbean to Bangladesh.

While early warning systems represent one type of intervention in this category, programmes to raise awareness of climate impacts and adaptation strategies is another. A good example comes from Zaragoza city, Spain. A city of 700,000, Zaragoza is located in a semi-arid region and receives annual precipitation of only 314 millimetres. Climate change is set to increase the number of consecutive dry days across Europe, so water shortages across the region may worsen. A report by Climate-ADAPT (2016) charts the response the city government, involving a structured and strategic programme to encourage the more judicious use of water and raise awareness to curb water demand. Conceptualised in four phases in collaboration with a local non-governmental organisation (NGO), the initiative started with a campaign to reduce water consumption within homes, public buildings and commercial activity through behavioural change and water-saving technology. The second phase entailed the demonstration of 50 water-saving technologies in parks, gardens and public buildings to encourage their wider uptake. Following this, the focus was on the city's major water-consuming actors, and guides and toolkits on reducing water use were distributed. Finally, the initiative worked to collect public commitments to reduce water use from citizens and businesses. As a result of this initiative, the city reduced water consumption from 136 litres per capita per day in 2000 to just under 100 litres per capita per day in 2010, exceeding its own targets. Crucially, there was a big jump in the number of people aware of water-saving techniques as a result of the programme.

3.1.3 Governance for community resilience

Across the world, government and non-governmental actors are establishing novel institutional arrangements at the local level to mount an effective response to the impacts of climate change in cities. One good example of this is the Surat Climate Change Trust in India. A city of 4.4 million, Surat is a vitally important economic centre in the western Indian state of Gujarat. The city suffers from numerous climate risks including floods, extreme rainfall, rising temperatures, sea level rise, storm surges and water scarcity (Patel 2009). In response, the city established the Surat Climate Change Trust, a registered institution led by Surat Municipal Corporation and comprised of representatives from local academic institutions, government departments, civil society organisations, technical bodies, the urban local body and chambers of commerce and industry. The trust is charged with leading resilience planning processes for the city, providing a sustainable and neutral platform to bring key institutions together to collaboratively identify solutions to complex urban problems, build the capacity of all relevant actors to enhance urban resilience and spread awareness on climate impacts and adaptation strategies. Since its establishment in 2010, the trust has acted as the focal point for all actions to enhance resilience in Surat. As part of this, it has steered processes to consolidate climate resilience strategies for Surat through wide and deep consultation with a variety of stakeholders. The trust has also undertaken numerous pilot projects, such as a flood early warning system, a research centre for investigating the health impacts of climate change and a mobile-phone-based system to monitor the delivery of basic services (Yadavar 2017). In this way, the trust provides a useful illustration of the way innovating governance arrangements and institutions can bring together diverse stakeholders to reduce the vulnerability of urban centres to climate change impacts.

Another set of activities that fall into this category is co-management and collaborative governance of natural resources. This is "a process of management in which government shares power with resource users, with each given specific rights and responsibilities relating to information and decision-making" (OECD 2001). This typically involves using community networks and, at times, establishing novel institutional arrangements at community level to manage natural resources. This approach has proved successful in the management of common property resources that have a crucial bearing on the adaptive capacity of vulnerable communities (Poricha and Dasgupta 2011).

This approach is illustrated in Delwara, a small peri-urban township in Rajasthan, India. The town suffers from acute water shortage, set to worsen with climate change (Khan 2019). A participatory needs assessment revealed that one of the reasons for this was that the main water body that had historically been the primary water source for the town had dried up due to "insufficient rainfall and because of the dilapidated condition of the pond feeders, the wells and aquifers of the settlement" (Poricha and Dasgupta 2011: 281). Spurred on by a local civil society organisation, a group of concerned citizens decided to remedy the situation by coming together to desilt the pond, repair feeder channels and rehabilitate aquifers. This has led to rejuvenation of the pond, that now has water throughout the year and has helped mitigate the risk of water shortages. The rejuvenation of this community water body has had positive knock-on effects on the water table and helped recharge surrounding water bodies as well (Poricha and Dasgupta 2011).

3.1.4 Resilient infrastructure

As noted throughout this book, infrastructure-based interventions have been a primary approach in resilience building, including at community level. This category of interventions includes "infrastructure for physical hazard defence, either based on engineering efforts (e.g. sea walls) or utilising ecological properties for protection (e.g. the replanting of hill sides to prevent mud slides)" (Paterson and Charles 2019: 333).

Projects that aim to provide nature-based solutions for reducing hazards and increasing resilience in cities are examples of such approaches. Chennai, India, a city of over 7 million and one the main urban centres of Southern India (Government of India 2011) has suffered in the recent past from hazards including catastrophic flood as well as a crippling water crisis (Ge 2019; Guntoju et al. 2019). There are multiple causes for this, one being that the water bodies and wetlands that have historically delivered vital ecosystem services by acting as flood buffers and recharging groundwater have depleted and deteriorated (Chandrasekhar 2017). This is why a group of civil society organisations in collaboration with local citizens and academic institutions is undertaking an initiative to restore the 100-acre

Sembakkam lake, that has been severely degraded due to excessive silt accumulation, disposal of untreated sewage and indiscriminate dumping of solid waste (The Nature Conservancy 2020). This resulted in the lake having diminished carrying capacity (and, thus, reduced ability to regulate flooding), and neighbourhoods around the lake are unable to use the water for drinking. The project is working with local communities to desilt the lake, spread awareness on pollution, encourage proper waste management, undertake clean-up drives, eradicate the flow of untreated sewage, clean inflow and outflow canals and strengthen bunds (Swaminathan 2018; Viswanathan 2019). In this way, the project hopes to utilise the ecological properties of the lake to protect the city from the most pressing impacts of a changing climate.

Another example of this category of initiative is a community-based water harvesting initiative in Semarang, Indonesia. Semarang is the capital and, with a population of 1.8 million, the largest city of the Java region (World Atlas 2020). A climate vulnerability assessment revealed drought and water scarcity to be major problems in the city that are likely to be exacerbated by climate change (Mercy Corps 2010). Moreover, for a number of reasons, including the topography of the city, the water utility is unable to service large parts of the city, and as a result, 60% of the city's population is not connected to water supply and is forced to rely on shallow wells that are not dependable (United Nations Climate Change 2020a). To reduce the vulnerability of urban residents to water scarcity, a group of civil society organisations in collaboration with local communities and international partners initiated a pilot project on rainwater harvesting (ADB 2014). Harvesting rainwater permits underground aquifers to recharge, ensuring the health of wells; it also controls runoff, which reduces flooding (United Nations Climate Change 2020a). As part of the pilot project, participatory meetings were held in select neighbourhoods with affected communities to identify individual as well as communal sites for the installation of harvesting infrastructure (Syam 2015). Following this, in collaboration with the community, rainwater harvesting infrastructure was installed in five homes and a neighbourhood school (United Nations Climate Change 2020a). Given the scale of this initiative, the flood prevention benefits are not apparent, but the community has reported a much more stable supply of underground water (Syam 2015). Seeing the success of this pilot, the Semarang Environmental Agency has decided to expand it to another 49 locations.

3.2 CHALLENGES FOR ENHANCING URBAN COMMUNITY RESILIENCE

Our review of the variety of existing approaches for enhancing urban resilience at community level reveals a series of key challenges. This section takes a critical lens to existing practice, arguing that much community-based action fails to engage with the structural drivers of risk and resilience, is not sensitive to scale, is prone to the pitfalls of participatory initiatives and may not be scalable or sustainable.

3.2.1 Lack of engagement with the structural drivers of risk

Following the findings of early pioneers of the study of political economy drivers of disaster risk and resilience, there is widespread understanding of the way that climate change impacts are mediated by structural conditions (Blaikie and Brookfield 1987; Blaikie et al. 2014). This may include discrimination (due to caste, language group, race or ethnicity), marginalisation (due to religious, gender, cultural or economic factors) or persecution (due to political affiliation or geographical or historical factors) (Broto 2017; Chu and Michael 2019). These factors can determine exposure to hazards; for instance, it is usually the economically and socially marginalised urban populations that reside on flood-prone land or hazardous slopes. These factors also influence coping and adaptive capacities, as they mediate the quality and extent of assets that households can access.

For instance, social or economic marginalisation leads to poor access to health and education services, which in turn affect career and income prospects and, thus, have a bearing on financial safety nets that a household can fall back on in times of crisis. Moreover, the very pattern of economic production and urban development in some cities can render certain communities less resilient than others (Dodman and Mitlin 2013). In the context of the COVID-19 pandemic too, these structural issues drive risk. This is evident in the early mortality rates in the United States: the virus caused the death of 1 in every 1,000 black Americans compared to 1 in every 2,150 white Americans (Scott and Animashaun 2020). Similarly, there is widespread acknowledgement of the gender inequality present in vulnerability to climate change, which reflects but also reinforces pre-existing gender inequalities (Eastin 2018).

Behind these structural factors driving risk are issues of power, empowerment and agency (Borie et al. 2019). Resilience initiatives that ignore such issues can be critiqued for merely engaging with the "proximate" causes of risk and harbouring an incremental vision of change (Pelling et al. 2015). Dodman and Mitlin note that "there is little point in asking local residents to participate in a project to landfill their site to reduce the risk of flooding if, at the same time, they are facing eviction due to lack of legal tenure" (2013: 646). Ziervogel et al. highlight how dominant approaches focus on the technical and emphasise infrastructure as opposed to rights, entitlements and governance. They argue that

> if the goal of resilience planning is to support risk management as well as just processes and outcomes of development, then it is not the pipes and roads of city infrastructure that need to be resilient. Rather, it is the rights and entitlements of urban citizens.
>
> (Ziervogel et al. 2017: 124)

The dominant methods for enhancing the resilience of urban communities promote an incremental approach that frequently fails to recognise the structural drivers of vulnerability (Pelling et al. 2015, Broto 2017). Examples discussed in Section 3.1 illustrate this critique, such as the community-based flood early warning initiatives failing to engage with the underlying reasons for certain communities occupying flood-prone land in the first place, focusing instead on saving lives and assets from flood events. Similarly, the initiative to restore Sembakkam lake in Chennai does not tackle the underlying drivers of risk (i.e. rampant urban expansion, inadequate enforcement of building norms or inadequate waste management infrastructure), instead choosing to focus on more proximate issues.

3.2.2 Challenges of working across scales and at scale

Linked to the lack of engagement with structural drivers of risk is the fact that most existing approaches concentrate their efforts at the local level. This is despite wide recognition that risk factors may extend beyond the local level and that enhancing resilience may require engagement across scales (Broto 2017). Institutional, legal and economic regulations determined at provincial or national levels may put people in cities at risk (Forsyth 2013). For instance, land use planning regulations may be set by central or federal ministries; these regulations have a strong influence on the location of new settlements or the quality of urban development, factors that can have a major impact on the exposure and vulnerability of urban residents to shocks and stresses. Similarly, macroeconomic development regimes set by governments at higher scales can have a substantial impact on equality and income distribution, which in turn influences vulnerability. Moreover, geographically speaking, hazards may be generated by areas that lie beyond the local, as illustrated by the influence of upstream activity on urban flood risk.

Solutions for reducing vulnerability and enhancing resilience require the engagement of actors and institutions across different scales (Park et al. 2012). For instance, in most countries the army and national disaster response teams are commanded by the provincial or federal governments and these can play a major role in supporting communities to bounce back after disaster events and prevent catastrophic damage. Building resilience over the long term requires shifts in institutional arrangements and policy reform, which can have substantial financial implications. This is difficult in towns and cities across the Global South as urban areas often do not have the authority to effect these changes and alter budget allocations (Bahadur 2014). Urban centres in countries such as India and Bangladesh are afflicted by fractured processes of decentralisation, where power to make administrative decisions is not matched by devolved power over finances and policy, which is retained at higher scales (Bahadur and Thornton 2015).

This holds true for processes of responding to COVID-19. While emergency services, usually controlled by city or sub-national governments, have led the response to the pandemic around the world, ensuring that further outbreaks are prevented will require shifts in institutional practice and changes in policy and finance matters where cities have more limited power. As such, the failure of community resilience-building initiatives to forge links with political structures above the most immediate political authority limits their ability to address urban resilience comprehensively and sustainably (Dodman and Mitlin 2013).

The example of co-managing an urban waterbody in Dewara, Rajasthan, reflected this challenge, as the actions of the individuals and organisations engaged were focused at the local level. A more targeted engagement with city and provincial authorities may have resulted in policy and institutional shifts that could have helped ensure enhanced sustainability and potentially call attention to other similar water bodies perishing across urban centres in the region.

This apart, most of interventions described in Section 3.1 are pilots, demonstration projects or small experiments that do not deliver impacts over a large geographical area or cover large proportions of the local population (Pal et al. 2019). The typhoon-resilient housing initiative covered only 244 houses in a city of 1.23 million, and the peri-urban agriculture initiative involved only 47 farmers and covered a mere 43 acres. Similarly, the rainwater harvesting initiative in Semarang covered only five households and one school, and even when scaled up by the authorities, the initiative was replicated in only another 49 locations in a city of 1.28 million. The microfinance project also identified itself very clearly as a pilot intervention. These initiatives are only "examples" of projects aimed at enhancing the resilience of urban communities; nonetheless, they are representative of the broader field. Commenting on the reason for this, Stead (2016: 40) notes that

> experimentation has gained currency as a means to deal with the complexity and uncertainty inherent to climate change issues because it allows for initiatives which have tangible and measurable results but they do not need to be conclusive to be persuasive and convincing.

Another reason for this is that a majority of these initiatives take place within the paradigm of international development cooperation and are conceived as "projects" with finite budgets and short timelines that severely limit their scope (Park et al. 2012). The relatively small scale at which most of these initiatives are implemented means that they are unable to effect change at the city scale and beyond, nor present models that are replicable beyond the limited contexts in which they have been implemented (Forsyth 2013).

3.2.3 Drawbacks of participatory techniques

Most methods for enhancing community resilience employ participatory approaches and methods for engagement, including focus group discussions, participatory mapping and transect walks, shared learning dialogues, collaborative exercises for prioritising adaptation options , self-report surveys, historical diagramming, games and role play (McCracken et al. 1988; Reed et al. 2013). All examples described in Section 3.1 employed these in some way. For instance, the initiative for harvesting rainwater in Semarang drew on shared learning dialogues with urban residents to acquire a bottom-up perspective on climate risk, and the initiative to restore Sembakkam lake relied on focus group discussions with neighbouring communities to examine the causes of environmental degradation (Mercy Corps 2010; Swaminathan 2018).

Despite the benefits of these approaches, they are not without their problems. They can end up excluding the most vulnerable constituents of urban communities through elite capture (Lund and Saito-Jensen 2013; Chu and Michael 2019). Conversely, those particularly at risk can be excluded from these methods due to their socio-economic or legal status. For instance, poor migrants often occupy the most exposed urban land and have high vulnerability but, due to lack of land tenure and lack of political voice, may not be considered citizens that need to be consulted (Chu and Michael 2019).

Those leading engagements with communities using these approaches need to be properly trained to ensure that they are mindful of their own positionality and how this might influence the voices that are included. In one typical case, a highly vulnerable "low caste" section of an urban flood-affected community was excluded from participatory exercises to inform resilience-building measures because those leading the exercises were from a higher caste and unfamiliar with this community (Bahadur 2014). Participatory approaches also assume a degree of social cohesion at the community level that is usually absent in urban contexts. This is because towns and cities are socially dynamic spaces with rapid cycles of migration that can lead to people from diverse ethnic, geographic and cultural backgrounds settling in particular neighbourhoods. On top of this, enhancing resilience to shocks and stresses requires shifts in behaviour, and influencing these shifts through communication and persuasion initiatives requires an understanding of values, which can differ radically in densely packed urban areas. This leads to fractured compliance with potentially significant measures to reduce risk. This has been seen in the response to the COVID-19 pandemic where mandates for self-isolation were not followed uniformly across large urban centres.

3.2.4 Sustainability and replicability

Another critique of current approaches to enhancing resilience of urban communities concerns the degree to which their benefits will continue to be felt after the direct involvement of an implementing agency or organisation ceases. A number of studies and reviews have found negative perceptions of the long-term sustainability of projects and continuing dependence on outside aid to continue projects, that hinders successful future community adaptation (Clissold et al. 2018; Clarke et al. 2019). Kernaghen and Silva (2014), speaking in the context of urban centres in Asia, examined the reasons for this lack of sustainability; they argue that donor-funded urban resilience and adaptation initiatives have traditionally suffered from finite budgets and short timescales (especially in comparison to government-led urban development initiatives), which in turn impacts the durability of outcomes and benefits that they deliver.

Replication and scaling up are also challenges, with solutions not necessarily applicable beyond the narrow contexts in which they are implemented and community-based initiatives not necessarily relevant to risks and policies that exist beyond the communities in which they are unfolding (Forsyth 2013). Ayers and Forsyth (2009: 29) argue that there exist

> several examples of successful projects at the scale of a household, a village, or a collection of villages. But do these local, community-based initiatives offer lessons for how to adapt to climate change at the national scale, or in other countries? The contextual nature of community-based adaptation makes developing indicators and models problematic, risking the proliferation of a piecemeal approach that lacks clarity and fails to attract wider climate change and development investment.

These issues are also evident in the examples of community-based initiatives explored in Section 3.1. For instance, the microfinance initiative for enhancing adaptation and resilience in Latin America required financing and coordination from a consortium of donors and implementing organisations to keep it going. Similarly, the multi-stakeholder climate change trust is a model that is relevant for the local context of Surat city, but the approach has not been replicated in other parts of the province or country. As a consequence of these experiences, critics question the degree to which dominant approaches for enhancing the resilience of urban communities are sustainable and replicable.

3.3 PIVOTING TO TRANSFORMATION IN COMMUNITY-BASED URBAN RESILIENCE

Despite the popularity of community-based approaches to supporting urban resilience, they face significant challenges, as set out above. They fail to engage with the underlying structural conditions that enhance risk or to include the voices of the most marginalised, and we have previously critiqued these approaches to urban resilience as taking a techno-managerial and instrumental approach to issues that are intrinsically about power and politics (Bahadur and Tanner 2014b). At the same time, the difficulties of engaging beyond the local level and the challenges of scalability of initiatives tend to support incremental approaches that tweak, rather than transform, the risk profiles of urban areas. Our proposed pivot aims to move resilience building beyond such apolitical and incremental approaches towards those that can be considered more transformational.

The concept of transformation has been explored in diverse disciplinary silos. Within the study of social-ecological systems, transformation occurs when a system flips from one state to another, and transformability is understood as "the capacity to create a fundamentally new system … when ecological, economic, and/or social conditions make the existing system untenable" (Walker and Salt 2012: 165). Transformation is also conceptualised in the literature on socio-technical transitions, where it is understood as shifts in "configurations of institutions, techniques and artefacts, as well as rules, practices and networks that determine the 'normal' development and use of technologies" (Smith et al. 2005: 1493). Within education, transformative learning has been conceptualised as going beyond knowledge acquisition to instead support critical awareness of the ways that learners learn and make meaning of their lives (Simsek 2012). This aspect of conscious transformation has been applied in the context of developing and claiming rights that address power imbalances in society in order to reduce vulnerability and enhance resilience (Devereux and Sabates-Wheeler 2004; Pelling 2010).

We draw on this growing range of conceptualisations and frameworks of transformation, particularly those relating to tackling climate change and disasters risk, to distil a set of key areas where attention is required.

3.3.1 Addressing the root causes of risk

Many conceptualisations of transformational adaptation and resilience underline the importance of addressing the root causes of vulnerability and risk (Pelling et al. 2015; Lonsdale et al. 2015; Patterson et al. 2018). Addressing these root causes or structural drivers of risk necessitates engaging with issues of power, justice and agency (Dodman and Mitlin 2013; Forsyth 2013; Pelling et al. 2015; Wolfram 2016). At the personal level, resilience initiatives that are aiming to deliver transformational outcomes could enable a process of "conscientisation", a critical awareness where individuals begin to recognise the structures of power that influence them and mediate the degree to which they are at risk (Kapoor 2007; Bivens et al. 2009; Pelling 2010). For instance, women may unproblematically accept tasks that put them at greater risk of climate impacts (e.g. fetching water for the household over long distances in the face of increasing heatwaves), or those belonging to historically marginalised social groups, such as lower castes in India, may not question why they occupy the most exposed land in any locality.

Beyond the personal, transformational approaches may also entail novel governance or institutional arrangements that enable communities to have greater agency and voice in decision-making (Few et al. 2017; O'Brien 2018). Existing development processes distribute risk unevenly, and risk management that changes this distribution will have implications for equity (Friend et al. 2016). Extending voice and agency to the vulnerable in these processes of decision-making may deliver structural improvements in their ability to manage risks and become resilient (Ziervogel 2019). Issues of power, justice and agency are particularly important for urban areas in developing countries due to rapid rates of in-migration; those from poorer rural populations are more likely to lack a voice and face political marginalisation in host communities, which has a direct impact on their vulnerability and exposure. A focus on issues of power and agency present within transformational approaches can also help mitigate the risk that participatory approaches will exclude the voices of vulnerable and marginalised people. An urban resilience initiative that unfolded in Gorakhpur, India, took these insights on board to tackle the problem of urban flooding (see Box 3.2) (Bahadur and Tanner 2014b).

3.3.2 Delivering lasting change

Mustelin and Handmer (2013) differentiate between incremental and transformational change and argue that the former focuses on short-term change, whereas the latter describes change that persists over long periods of time. This is borne out in applied approaches, where for an initiative to be transformational, it needs to deliver outcomes whose benefits will endure over the long term (Bahadur et al. 2015; Lonsdale et al. 2015). The United Kingdom's support to developing countries through International Climate Finance prioritises sustainability as an essential parameter of transformational change and says change must be sustained even after direct funding stops (DFID 2014).

Pal et al. (2019: 9) reinforce the critical importance of sustainability to their conceptualisation of transformation but underline that this

> can take many forms: it could include uptake of models and approaches into public policies and programmes; the establishment of permanent governance structures that will continue to deliver benefits; changing attitudes and behaviours; or a permanent improvement in the processes through which public organisations can have a critical role in reducing risk and vulnerability.

Both Few et al. (2017) and O'Brien (2018) emphasise that such lasting change may result from a reorganisation of government systems and structures or from a reform of political and administrative architectures. This issue of sustainability is particularly salient because governance mechanisms and structures tend to be concentrated in urban areas, thereby offering an opportunity to embed resilience-enhancing policies and institutions to deliver lasting changes.

This vision of transformation as sustainable change can be illustrated in numerous ways. An important interpretation of this idea can be seen in the establishment of new government institutions charged with enhancing the resilience of urban communities. The New York City Mayor's Office of Recovery

BOX 3.2 Putting people, power and politics at the heart of urban resilience in Gorakhpur, India

Many parts of Gorakhpur are affected by high-frequency low-level flooding (locally known as waterlogging) that disrupts life and causes severe damage to people's health and property. Climate change is set to exacerbate the problem (Kannan 2009; Rockefeller Foundation 2010). The proximate cause of flooding is poor drainage; the main drains of the city have not been upgraded since the colonial period and are clogged with garbage. Instead of opting to invest scarce resources only in cleaning and renovating drains, an urban resilience project attempted a more systemic and structural solution (GEAG 2009).

First, those running the project initiated wide and deep engagement with affected communities to generate a shared understanding of the causes of and solutions to the problem (Reed et al. 2013). A key issue that came up in the Maheva neighbourhood, which had been particularly badly impacted, was the lack of municipal services such as solid waste management (that led to the clogging of drains), public works (that led to the drains falling into disrepair) and community health (that led to waterborne diseases) (Bahadur 2014). This was partly because the neighbourhood was controlled by a powerful local municipal councillor who did not acknowledge the need to prioritise these services (Bahadur and Tanner 2014b).

The project organised community members into teams of volunteers who started to gradually challenge his control. Volunteers fanned out through the neighbourhood and spread awareness of the issue, and this in-turn increased demands for municipal services and put pressure on the councillor to be accountable. As part of their actions to raise awareness, volunteers helped convene large community meetings to discuss problems that the community was facing, with a view to finding solutions to a range of civic problems, thereby contributing to conscientisation. Volunteers also started to destabilise the source of the councillor's power by harming their material interests. For instance, Maheva's lack of adequate solid waste management was addressed by volunteers through new arrangements for garbage collection and disposal that threatened the existing, malfunctioning system of waste management, that was allegedly a source of kickbacks.

Partly as a result of this, in the subsequent election the councillor was replaced with someone more sensitive to this issue. At a higher scale, the project lobbied for and established a platform where citizens and representatives of affected communities could engage directly with the leadership of the urban local body, allowing them a voice in crucial decision-making processes to which they had no access previously (Bahadur and Tanner 2014a). The project also engaged with the provincial and central governments to promote a favourable policy environment at the local level. This led to improvements in the extension of municipal services to affected communities. In this way, the initiative "started in a tangible way to challenge political power in the local communities and to embody adaptation as transformation through the generation of new rights claims" (Bahadur and Tanner 2014b: 210).

and Resiliency (ORR) was established in 2014 in the aftermath of Hurricane Sandy and leads a USD 20 billion resilience programme aiming to make New York City more resilient to the impacts of climate change (USDoE 2018). The ORR works to ameliorate the impact of coastal storm surges, sea level rise, heatwaves and flooding – the main impacts of climate change that are set to affect New York city in the coming decades. It is staffed by a chief resilience officer and a team of resilience experts that are a formal part of the mayor's office and on the government payroll. The ORR is charged with executing a strategic vision for the city's resilience, that has been enshrined in the OneNYC plan (Mesa 2015). This includes a major thrust on ensuring the resilience of communities through citizen–government collaboration on strengthening community-based institutions that can support risk reduction and recovery actions, enhancing social capital at the neighbourhood level and establishing "social empowerment" zones for distressed communities where the government allocates additional time and money to improve community cohesion.

The ORR is a permanent institution charged with improving emergency preparedness and planning, ensuring business continuity during disturbances, making sure that dividends of resilience are distributed equitably, preparing the city for heat extremes, overseeing that buildings in the city are structurally resilient, working with public agencies and the private sector to adapt infrastructure to the expected and uncertain impacts of climate change, and shoring up coastal defence against storms and sea level rise (Mesa 2015). The ORR is an example of an innovative governance structure for resilience that is permanent, working to a long-term plan and rooted in engagement at the community level. By programming vast resources to enhance resilience and overseeing the execution of an expansive resilience strategy, the ORR is transforming the way New York City and its residents engage with climate change risk.

3.3.3 Working at scale

Different conceptualisations of transformation also underline the critical importance of scale; both working *at* scale (i.e. magnitude) and *across* scales (i.e. spanning tiers of governance). In their review of transformational adaptation, Kates et al. (2012: 7157) argue that "common adaptations can become transformational when they are used at a greater scale or in integrated combinations with much larger effects than before". Few et al. (2017: 4) count "expansion" as a key attribute of transformation and argue transformational change can result from resilience-building measures being implemented at great scale and intensity. The United Kingdom's International Climate Finance has also emphasised that scale can help interventions achieve transformation through institutional and policy reform or by driving down the costs of technology (DFID 2014). Reflecting its funding though International Climate Finance, the BRACED initiative included scale as one of its three pillars of transformation (Villanueva et al. 2015).

While a focus on the local scale is key, transformative capacity requires work across scales, spanning "individuals, households, groups, organizations, networks as well as society at large" and administrative boundaries (e.g. districts, metropolitan areas and the national level) (Dodman and Mitlin 2013; Wolfram 2016: 128). This issue of working across scales is particularly important for urban areas in developing countries, because towns and cities here are afflicted by fractured decentralisation and the power to make key decisions for enhancing resilience may reside with authorities at the provincial and federal levels (Mukhopadhyaya et al. 2000).

The importance of scale is illustrated by the work of Slum/Shack Dwellers International (SDI), a network of community-based organisations of the urban poor in 32 countries, covering hundreds of cities and towns across Africa, Asia and Latin America (SDI 2016). At the heart of SDI's approach are community savings groups at the level of the informal settlement or neighbourhood, where households contribute a daily amount into a community-managed fund that can be used for a range of activities including those that can help reduce risk, such as structural improvements to dwellings, installation of basic services, or safety nets in times of crisis (d'Cruz and Mitlin 2005). SDI links savings groups to one another within cities, links federations within cities to those in other cities and then links countries within the network to one another. This structure permits the initiative to be firmly rooted at the local level but extends to it a global scale that allows for greater impact.

Individually, each community savings group does not necessarily generate finances large enough to upgrade settlements to a point where their level of risk is transformed for the better, but by pooling resources between neighbourhoods, cities and countries, the network is able to mobilise substantial finances (Smith et al. 2014). The Urban Poor Fund International, a mechanism that capitalises local funds established by the 33 national federations across 464 cities within the SDI network, has mobilised USD 20 million. This has "improved the living conditions of more than 200,000 poor people in informal and low-income settlements through secured land tenure, improved infrastructure and basic services" – actions that have a substantial bearing on community resilience (Smith et al. 2014: 17).

Apart from finances, working across scales in the form of a federation that operates at the city level and nationally also provides bargaining power and policy influence to the residents of informal settlements who have been historically marginalised (d'Cruz and Mitlin 2005). This policy influence has been utilised in many ways, most prominently to avoid eviction of slum dwellers, challenge their relocation to worse areas (that could lead to heightened climate risk exposure), secure structural improvements and enhancements (that reduce risk) and ensure the supply of basic services (that reduce vulnerability) (Khan 2014). SDI also leverages the scale of its network to enable person-to-person exchanges between its various federations, which helps cross-pollinate ideas, engender innovation and consolidate the global footprint of the organisation that has local communities at its core (Khan 2014). Individually, community groups are less likely to be effective, but when organised into a transnational network, they have demonstrated the potential to mobilise financing and political power to deliver transformational outcomes.

3.3.4 Inducing catalytic change

An important drawback of dominant approaches for building the resilience of urban communities is that they could be seen as localised experiments that are not replicable beyond the specific context in which they unfold, and therefore have a limited impact (Forsyth 2013). There is a growing understanding of how transformational approaches entail cascading impacts and influence change beyond the direct area in which they are implemented (Kates et al. 2012).

Pal et al. (2019: 8) argue that "transformational change can result from initiatives that *catalyse* broader change. Deliberate shifts within systems can be expanded to trigger indirect changes and cascading impacts within structures and systems that are beyond an initiative's direct mandate or reach". The BRACED initiative underlines this point to argue that catalytic impact could entail the replication of particular adaptation measures in another geographic area; the autonomous transmission of learning or skills from the beneficiaries of a project to another group outside the project's purview; and the use by multiple agencies of a model, tool or framework developed by one initiative, delivering benefits for vulnerability reduction in areas beyond the original remit (Villanueva et al. 2018).

The Self Employed Women's Association (SEWA) in India illustrates an approach that has delivered such catalytic change. Established in 1972 as a trade union for urban working women, the organisation aims to build the capacity of women through education and training, catalyse the formation of cooperatives and producer groups and facilitate women's access to social security and social protection (de Luca et al. 2013). At the heart of SEWA's model is the establishment of Self Help Groups (SHGs) that allow community members to band together to pool resources and take out loans for various reasons, including livelihood activities and meeting emergency expenses. SHGs are also a source of social support, technical help and advice for its members.

Crucially, SHGs play an important role in enhancing the resilience of poor and vulnerable households (Weingärtner et al. 2017). Productive loans from SHGs can help improve households' financial status, providing a vital safety net for emergencies; or more directly, as in SEWA's case, they can provide loans to repair and improve housing after disasters, allowing communities to bounce back. The SHGs also provide "life skills" training in areas such as literacy, health and hygiene, and financial management or in specific technical/vocational skills that strengthen lives and livelihoods of households, increasing their overall capacity to weather shocks and stresses. In addition, SHGs build social capital, that is a vital component of resilience as "it can instil a sense of mutual reciprocity, facilitate recovery, and strengthen networks to spread information to enhance preparedness" (Weingärtner et al. 2017: 13).

Apart from its horizontal influence, spreading from one initial city in the state of Gujarat to towns and villages across 12 Indian states, this model has also catalysed change vertically by influencing national policy. Based on the success of the SHG model, the Government of India adopted it as the lynchpin of the National Urban Livelihood Mission. Established in 2013, the mission envisages that at least one member from every urban poor household in each of the 4,041 towns and cities in India will be included in an SHG established by the mission (MoHUA 2018; PIB Delhi 2019). Each SHG will have 10–20 members and will, by and large, play the same function as those established under SEWA. Up until 2016, 454,000 urban poor people have received training and 73,476 individuals have accessed over USD 75 million in loans (PTI 2016). The scheme was given an outlay of USD 70 million from the national budget in the financial year 2019–2020 (PTI 2019). Thus, this model of community development (that delivers resilience benefits) pioneered by an NGO, catalysed a national programme that is being rolled out at scale.

3.4 CONCLUSION

This chapter began with a review of the state of play in building community resilience in urban centres by examining approaches that aim to enhance resources, raise awareness, improve governance and build community-level infrastructure. It then went on to examine how these existing approaches fail to engage with the structural drivers of risk, overlook the importance of engaging tiers of government beyond the local, deliver results on a small scale and result in improvements that are not durable. Overall, the current of state of play delivers a vision of change that is incremental. This in turn creates a need for more transformational approaches that engage with the underlying drivers of risk and work at and across scales to deliver sustainable outcomes that catalyse broader shifts.

In conclusion, a pivot from incremental to transformational approaches will entail key stakeholders being challenged to recognise the potential of this new way of delivering community resilience. Supporting international partners need to be challenged to shift away from commissioning short-term interventions that are less likely to engage with political structures, as these result in instrumental approaches for engaging with structural problems. Civil society organisations working on issues of climate and development at the local level should be urged to develop networks and federate with other organisations doing similar work in order to influence governance structures at the provincial and national levels. Those leading change processes need to ensure that the theories of change they follow reflect an appreciation of structural barriers for resilience and the need to deliver sustainable results at scale. Evaluation experts need to be pushed to build on emerging insights on methods of tracking and gauging transformational shifts to develop operational approaches for monitoring and evaluating transformation.

Finally, as transformation entails a departure from established processes and protocols, it portends disruption. Change of this nature will get noticed and upset the status quo, and it is therefore likely to face resistance. Those heralding such change need to be politically smart and act as policy entrepreneurs to ensure that the right stakeholders are on board and to exploit windows of opportunity (Tanner et al. 2019).

REFERENCES

Agrawala, S. and Carraro, M. (2010). *Assessing the Role of Microfinance in Fostering Adaptation to Climate Change*. OECD Environmental Working Paper No. 15. Paris: OECD.

Aldrich, D. P. and Meyer, M. A. (2015). Social capital and community resilience. *American Behavioral Scientist*, *59*(2), 254–269.

Archer, D., Almansi, F., DiGregorio, M., Roberts, D., Sharma, D. and Syam, D. (2014). Moving towards inclusive urban adaptation: Approaches to integrating community-based adaptation to climate change at city and national scale. *Climate and Development*, *6*(4), 345–356.

Asian Development Bank. (2014). *Urban Climate Change Resilience; A Synopsis*. Manila: Asian Development Bank.

Ayers, J. and Forsyth, T. (2009). Community-based adaptation to climate change. *Environment: Science and Policy for Sustainable Development*, *51*(4), 22–31.

Bahadur, A. (2014). *Policy Climate and Climate Policies* [PhD thesis]. University of Sussex, UK.

Bahadur, A. V. and Tanner, T. (2014a). Policy climates and climate policies: Analysing the politics of building urban climate change resilience. *Urban Climate*, *7*, 20–32.

Bahadur, A. and Tanner, T. (2014b). Transformational resilience thinking: Putting people, power and politics at the heart of urban climate resilience. *Environment and Urbanization*, *26*(1), 200–214.

Bahadur, A. V. and Thornton, H. (2015). Analysing urban resilience: A reality check for a fledgling canon. *International Journal of Urban Sustainable Development*, *7*(2), 196–212.

Bahadur, A. V., Peters, K., Wilkinson, E., Pichon, F., Gray, K. and Tanner, T. (2015). *The 3As: Tracking Resilience across BRACED*. London: Overseas Development Institute.

Bharwani, S. (2011). *What is a community-based adaptation project?* WeAdapt. Retrieved 23 November, 2020, from https://www.weadapt.org/knowledge-base/community-based-adaptation-network-cba-network/what-is-a-community-based-adaptation-project

Bivens, F., Moriarty, K. and Taylor, P. (2009). Transformative education and its potential for changing the lives of children in disempowering contexts. *IDS Bulletin*, *40*(1), 97–108.

Blaikie, P. M. and Brookfield, H. (Eds). (1987). *Land Degradation and Society*. York: Methuen.

Blaikie, P., Cannon, T., Davis, I. and Wisner, B. (2014). *At Risk: Natural Hazards, People's Vulnerability and Disasters*. Abingdon: Routledge.

Brecht, H., Deichmann, U. and Wang, H. G. (2013). *A Global Urban Risk Index*. Washington DC: The World Bank.

Borie, M., Ziervogel, G., Taylor, F., Millington, J., Sitas, R. and Pelling, M. (2019). Mapping (for) resilience across city scales: An opportunity to open-up conversations for more inclusive resilience policy? Environmental Science & Policy, 99, 1–9.

Broto, V. C. (2017). Urban governance and the politics of climate change. *World Development*, *93*, 1–15.

Chandrasekhar, K. (2017). *Wetlands in peril, Chennai*. Conservation India. Retrieved 23 November, 2020, from https://www.conservationindia.org/gallery/wetlands-in-peril-chennai

Clarke, T., McNamara, K. E., Clissold, R. and Nunn, P. D. (2019). Community-based adaptation to climate change: Lessons from Tanna Island, Vanuatu. *Island Studies Journal, 14*(1), 59–80.

Climate-ADAPT. (2016). *Zaragoza: Combining awareness raising and financial measures to enhance water efficiency*. Retrieved 23 November, 2020, from https://climate-adapt.eea.europa.eu/metadata/case-studies/zaragoza-combining-awareness-raising-and-financial-measures-to-enhance-water-efficiency

Climate Investment Funds. (2018). *Microfinance for Climate Adaptation: From Readiness to Resilience*. Washington DC: Climate Investment Funds.

Clissold, R., Tahlia, C., Priebbenow, B. and McNamara, K. (2018). *Panacea for the Pacific? Evaluating community-based climate change adaptation*. New Security Beat. Retrieved 23 November, 2020, from https://www.newsecuritybeat.org/2018/02/panacea-pacific-evaluating-community-based-climate-change-adaptation/

d'Cruz, C. and Mitlin, D. (2005). *Shack/Slum Dwellers International: One Experience of the Contribution of Membership Organizations to Pro-Poor Urban Development*. London: International Institute for Environment and Development.

de Luca, L., Sahy, H., Joshi, S. and Cortes, M. (2013). *Learning from Catalysts of Rural Transformation*. Geneva: International Labour Organization.

Department for International Development. (2014). *KPI 15 Guidance*. London: Department for International Development.

Devereux, S. and Sabates-Wheeler, R. (2004). *Transformative Social Protection*. Working Paper 232. Brighton: Institute of Development Studies.

Dodman, D. and Mitlin, D. (2013). Challenges for community-based adaptation: Discovering the potential for transformation. *Journal of International Development, 25*(5), 640–659.

Dubbeling, M. (2014). *Integrating Urban Agriculture and Forestry into Climate Change Action Plans: Lessons from Sri Lanka*. London: Climate and Development Knowledge Network.

Eastin, J. (2018). Climate change and gender equality in developing states. *World Development, 107*, 289–305.

Few, R., Morchain, D., Spear, D., Mensah, A. and Bendapudi, R. (2017). Transformation, adaptation and development: Relating concepts to practice. *Palgrave Communications, 3*(1), 1–9.

Firman, T. (2017). The urbanisation of Java, 2000–2010: Towards "the island of mega-urban regions". *Asian Population Studies, 13*(1), 50–66.

Forsyth, T. (2013). Community-based adaptation: A review of past and future challenges. *Wiley Interdisciplinary Reviews: Climate Change*, *4*(5), 439–446.

Friend, R. M., Anwar, N. H., Dixit, A., Hutanuwatr, K., Jayaraman, T., McGregor, J. A., Menon, M. R., Moench, M., Pelling, M. and Roberts, D. (2016). Re-imagining inclusive urban futures for transformation. *Current Opinion in Environmental Sustainability*, *20*, 67–72.

Garland, A. M. (Ed.). (2007). *Global Urban Poverty: Setting the Agenda*. Washington DC: Woodrow Wilson International Center for Scholars.

Ge, K. (2019). *Rivers Remember: Chennai Rains and the Shocking Truth of a Manmade Flood*. Leicestershire: Context Publications.

Gorakhpur Environmental Action Group. (2009). *Vulnerability Analysis*. Gorakhpur: Gorakhpur Environmental Action Group.

Government of India. (2011). *2011 Census Data*. Office of the Registrar General & Census Commissioner. Retrieved 23 November, 2020, from https://censusindia.gov.in/2011-Common/CensusData2011.html

Greenfieldboyce, N. (2007). *Study: 634 million people at risk from rising seas*. NPR. Retrieved 23 November, 2020, from https://www.npr.org/templates/story/story.php?storyId=9162438

Güneralp, B., Güneralp, İ. and Liu, Y. (2015). Changing global patterns of urban exposure to flood and drought hazards. *Global Environmental Change*, *31*, 217–225.

Guntoju, S., Alam, M. and Sikka, A. (2019). *Chennai water crisis: A wake-up call for Indian cities*. Down To Earth. Retrieved 23 November, 2020, from https://www.downtoearth.org.in/blog/water/chennai-water-crisis-a-wake-up-call-for-indian-cities-66024

Hammill, A., Matthew, R. and McCarter, E. (2008). Microfinance and climate change adaptation. *IDS Bulletin*, *39*(4), 113–122.

IFRC. (2018). *Flood Early Warning Early Action System*. Geneva: International Federation of Red Cross and Red Crescent Organisations.

Kannan, K. (2009). *Gorakhpur looks different from the climate change lens*. Oxfam India. Retrieved 23 November, 2020, from www.oxfamindia.org/blog/kkannan/gorakhpurlooksdifferentclimatechangelens%E2%80%A6

Kapoor, R. (2007). Transforming self and society: Plural paths to human emancipation. *Futures*, *5*(39), 475–486.

Kates, R. W., Travis, W. R. and Wilbanks, T. J. (2012). Transformational adaptation when incremental adaptations to climate change are insufficient. *Proceedings of the National Academy of Sciences, 109*(19), 7156–7161.

Kernaghen, S. and da Silva, J. (2014). Initiating and sustaining action: Experiences building resilience to climate change in Asian cities. Urban Climate, 7, 47–63.

Khan, F. (2014). Adaptation vs. development: Basic services for building resilience. *Development in Practice, 24*(4), 559–578.

Khan, S. (2019). *Western Rajasthan faces climate change perils*. Times of India. Retrieved 23 November, 2020, from http://timesofindia.indiatimes.com/articleshow/69656801.cms

Krummheuer, A. R., Gruening, C. and Jungfleisch, C. (2015). Microfinance for ecosystem-based adaptation (MEbA) in Peru and Colombia. *Enterprise Development & Microfinance, 26*(3), 274–291.

Lonsdale, K., Pringle, P. and Turner, B. (2015). *Transformative Adaptation: What It Is, Why It Matters & What Is Needed*. Oxford: UK Climate Impacts Programme.

Lund, J. F. and Saito-Jensen, M. (2013). Revisiting the issue of elite capture of participatory initiatives. *World Development, 46*, 104–112.

McCracken, J. A., Pretty, J. N. and Conway, G. R. (1988). *An Introduction to Rapid Rural Appraisal for Agricultural Development*. London: International Institute for Environment and Development.

Mercy Corps. (2010). *Semarang Vulnerability and Capacity Assessment*. Jakarta: Mercy Corps Indonesia.

Mesa, N. (2015). *One New York*. New York: The City of New York.

Ministry of Housing and Urban Affairs. (2018). *Social Mobilisation and Institution Development: Deendayal Antyodaya Yojana – National Urban Livelihoods Mission*. New Delhi: Ministry of Housing and Urban Affairs, Government of India.

Mohamed, L. S. and Gunasekera, J. (2014). Promoting urban agriculture as a climate change strategy in Kesbewa, Sri Lanka. *Urban Agriculture Magazine, 27*, 20–23.

Mukhopadhyaya, A., Jayal, N., Meenakshisundaram, S., Benjamin, S. and Vyasulu, V. (2000). *Decentralisation in India*. New Delhi: United Nations Development Program.

Mustelin, J. and Handmer, J. (2013). Triggering transformation: Managing resilience or invoking real change? In *Proceedings of Transformation in a Changing Climate* (pp. 24–32). Oslo: University of Oslo.

O'Brien, K. (2018). Is the 1.5°C target possible? Exploring the three spheres of transformation. *Current Opinion in Environmental Sustainability, 31*, 153–160.

Opitz-Stapleton, S. (2013). *Da Nang: Typhoon Intensity and Climate Change*. Boulder, CO: ISET.

Organisation for Economic Co-operation and Development. (2001). *Co-management*. Retrieved 23 November, 2020, from https://stats.oecd.org/glossary/detail.asp?ID=382

Pal, U., Bahadur, A. V., McConnell, J., Vaze, P., Kumar, P. and Acharya, S. (2019). *Unpacking Transformation: A Framework and Insights from Adaptation Mainstreaming*. Action on Climate Today Learning Paper. New Delhi: ACT.

Park, S. E., Marshall, N. A., Jakku, E., Dowd, A. M., Howden, S. M., Mendham, E. and Fleming, A. (2012). Informing adaptation responses to climate change through theories of transformation. *Global Environmental Change, 22*(1), 115–126.

Patel, T. (2009). *Funding for Adaptation to Climate Change: The Case of Surat* [master's thesis]. Massachusetts Institute of Technology.

Paterson, B. and Charles, A. (2019). Community-based responses to climate hazards: Typology and global analysis. *Climatic Change, 152*(3–4), 327–343.

Patterson, J. J., Thaler, T., Hoffmann, M., Hughes, S., Oels, A., Chu, E., Mert, A., Huitema, D., Burch, S. and Jordan, A. (2018). Political feasibility of 1.5°C societal transformations: The role of social justice. *Current Opinion in Environmental Sustainability, 31*, 1–9.

Pelling, M. (2010). *Adaptation to Climate Change: From Resilience to Transformation*. Abingdon: Routledge.

Pelling, M., O'Brien, K. and Matyas, D. (2015). Adaptation and transformation. *Climatic Change, 133*(1), 113–127.

Phong, T. V. (2013). *Lessons from Typhoon Nari*. Boulder, CO: ISET.

PIB Delhi. (2019). *Self help groups*. Press Information Bureau, Government of India. Retrieved 23 November, 2020, from https://pib.gov.in/PressReleaseIframePage.aspx?PRID=1578572

Poricha, B. and Dasgupta, B. (2011). Equity and access: Community-based water management in urban poor communities: An Indian case study. *WIT Transactions on Ecology and the Environment, 153*, 275–285.

PTI. (2016). *Meet to resolve bottlenecks in Antyodaya Yojana implementation*. The Economic Times. Retrieved 23 November, 2020, from https://economictimes.indiatimes.com/news/economy/policy/meet-to-resolve-bottlenecks-in-antyodaya-yojana-implementation/articleshow/52925936.cms

PTI. (2019). *Budgetary provisions for Housing and Urban Affairs Ministry hiked by 17%*. Live Mint. Retrieved 23 November,

2020, from https://www.livemint.com/budget/news/budgetary-provisions-for-housing-and-urban-affairs-ministry-hiked-by-17-1549029360913.html

Reed, S., Friend, R., Toan, V. C., Thinphanga, P., Sutarto, R. and Singh, D. (2013). "Shared learning" for building urban climate resilience – experiences from Asian cities. *Environment and Urbanization*, *25*(2), 393–412.

Rockefeller Foundation. (2010). *Asian Cities Climate Change Resilience Network*. New York: Rockefeller Foundation.

Scott, D. and Animashaun, C. (2020). *Covid-19's stunningly unequal death toll in America, in one chart*. VOX. Retrieved 23 November, 2020, from https://www.vox.com/coronavirus-covid19/2020/10/2/21496884/us-covid–19-deaths-by-race-black-white-americans

Slum Dwellers International. (2016). *About us*. Know Your City. Retrieved 23 November, 2020, from http://knowyourcity.info/who-is-sdi/about-us/

Sharma, A. (2017, 8 December). *Cities of the poor: A view on urban poverty in India*. Times of India. Retrieved 23 November, 2020, from https://timesofindia.indiatimes.com/blogs/in-the-name-of-development/cities-of-the-poor-a-view-on-urban-poverty-in-india/

Simsek, A. (2012). Transformational learning. In N. M. Seel (Ed.), *Encyclopaedia of the Sciences of Learning*. Springer: Boston.

Smith, A., Stirling, A. and Berkhout, F. (2005). The governance of sustainable socio-technical transitions. *Research Policy*, *34*(10), 1491–1510.

Smith, B., Brown, D. and Dodman, D. (2014). *Reconfiguring Urban Adaptation Finance*. London: International Institute for Environment and Development.

Stead, D. (2016). Key research themes on governance and sustainable urban mobility. *International Journal of Sustainable Transportation*, *10*(1), 40–48.

Susandi, A. and Tamamadin, M. (2017). Toward a dynamic disaster decision support system in Citarum river basin: FEWEAS Citarum. In *Proceedings of 2017 International Conference on Smart Cities, Automation & Intelligent Computing Systems (ICON-SONICS)* (pp. 6–10). Yogyakarta: IEEE.

Swaminathan, T. S. (2018). *Residents interact with IIT-M experts*. The Hindu. Retrieved 23 November, 2020, from https://www.thehindu.com/news/cities/chennai/residents-interact-with-iit-m-experts-for-restoration-of-sembakkam-lake/article24412383.ece

Syam, D. (2015, 2 February). *Rain water harvesting project in Semarang, Indonesia*. Asian Cities Climate Change Resilience Network. Retrieved November 23, 2020, from https://www.acccrn.net/resources/rain-water-harvesting-project-semarang-indonesia

Tanner, T., Zaman, R. U., Acharya, S., Gogoi, E. and Bahadur, A. (2019). Influencing resilience: The role of policy entrepreneurs in mainstreaming climate adaptation. *Disasters*, *43*(S3), S388–S411.

The Nature Conservancy. (2020). *Restoring urban wetlands*. Retrieved 23 November, 2020, from https://www.nature.org/en-us/about-us/where-we-work/india/stories-in-india/restoring-urban-wetlands/

Tran, P., Tran, T. H. and Tran, A. T. (2014). *Sheltering from a Gathering Storm: Typhoon Resilience in Vietnam*. Boulder, CO: ISET.

United Nations Department of Economic and Social Affairs. (2018). *68% of the world population projected to live in urban areas by 2050, says UN*. Retrieved 23 November, 2020, from https://www.un.org/development/desa/en/news/population/2018-revision-of-world-urbanization-prospects.html

United Nations Climate Change. (2020a). *Rainwater harvesting as an alternative source of clean water in Semarang City – Indonesia*. Retrieved 23 November, 2020, from https://unfccc.int/climate-action/momentum-for-change/activity-database/momentum-for-change-rain-water-harvesting-as-an-alternative-sources-of-clean-water-in-semarang-city

United Nations Climate Change. (2020b). *Vietnam: Building storm-resistance houses*. United Nations Climate Change. Retrieved 23 November, 2020, from https://unfccc.int/climate-action/momentum-for-change/lighthouse-activities/building-storm-resistant-houses

United States Department of Energy. (2018). *NYC Mayor's Office of Recovery and Resiliency*. Retrieved 23 November, 2020, from https://betterbuildingssolutioncenter.energy.gov/sites/default/files/attachments/NYC.pdf

Villanueva, P. S., Gould, C., Gregorowski, R., Bahadur, A. and Howes, L. (2015). *BRACED Programme Monitoring & Evaluation (M&E) Guidance Notes*. London: Overseas Development Institute.

Villanueva, P., Itt, R. and Sword-Daniels, V. (2018). *Routes to Resilience*. London: Overseas Development Institute.

Viswanathan, N. (2019). *Sewage threat to Sembakkam lake restoration*. The New Indian Express. Retrieved 23 November, 2020, from https://www.newindianexpress.com/cities/chennai/2019/may/06/sewage-threat-to-sembakkam-lake-restoration-1973130.html

Walker, B. and Salt, D. (2012). *Resilience Thinking: Sustaining Ecosystems and People in a Changing World*. Washington DC: Island Press.

Weingärtner, L., Pichon, F. and Simonet, C. (2017). *How Self-Help Groups Strengthen Resilience: A Study of Tearfund's Approach to Tackling Food Insecurity in Protracted Crises in Ethiopia*. London: Overseas Development Institute.

Wolfram, M. (2016). Conceptualizing urban transformative capacity: A framework for research and policy. *Cities*, *51*, 121–130.

World Atlas. (2020). *Biggest cities in Indonesia*. Retrieved 23 November, 2020, from https://www.worldatlas.com/articles/biggest-cities-in-indonesia.html

World Bank. (2019). *Population living in slums (% of urban population)*. Retrieved 23 November, 2020, from https://data.worldbank.org/indicator/EN.POP.SLUM.UR.ZS?view=chart

Yadavar, S. (2017, 14 December). *What Surat can teach other Indian cities about public health*. IndiaSpend. Retrieved 23 November, 2020, from https://www.indiaspend.com/what-surat-can-teach-other-indian-cities-about-public-health-11451/

Ziervogel, G. (2019). Building transformative capacity for adaptation planning and implementation that works for the urban poor: Insights from South Africa. *Ambio*, *48*(5), 494–506.

Ziervogel, G., Pelling, M., Cartwright, A., Chu, E., Deshpande, T., Harris, L., Hyams, K., Kaunda, J., Klaus, B., Michael, K. and Pasquini, L. (2017). Inserting rights and justice into urban resilience: A focus on everyday risk. *Environment and Urbanization*, *29*(1), 123–138.

Chapter 4

Urban planning for resilience: Embracing informality

URBAN planning has played a central role in efforts for enhancing urban climate and disaster resilience (Storbjörk and Uggla 2015). This has been reinforced by international frameworks, with the 2005 Hyogo Framework for Action to reduce disaster risk urging signatories to "incorporate disaster risk assessments into the urban planning" (UNISDR 2005: 12). The Sendai Framework for Disaster Risk Reduction that replaced the Hyogo Framework in 2015 prioritises "the mainstreaming of disaster risk assessments into land use policy development and implementation, including urban planning" (UNISDR 2015: 19). Similarly, Target 11B of the Sustainable Development Goals underlines the importance of urban policies and plans for adaptation to climate change and resilience to disasters (UN 2019), and the New Urban Agenda (adopted at the 2016 Habitat III Conference) highlights the importance of "mainstreaming holistic and data-informed disaster risk reduction and management" into urban planning processes (UN-Habitat 2017: 21).

A host of bilateral and multilateral organisations reinforce the norm of mainstreaming resilience into formal urban planning processes. For instance, policy guidance from the Asian Development Bank (2016: 2) notes how "urban land use management processes such as land use planning, development control, greenfield development, and urban redevelopment can play an important role in reducing disaster risk". Guidance from the World Bank (2015a: 21) underlines that "[b]uilding codes and land use regulation have a crucial … role to play in investment programs for reducing disaster and chronic risk". USAID's

(Cote and King 2017: viii) approach to urban resilience planning states that "resilience thinking needs to be mainstreamed into standard development frameworks and planning", while the International Federation of the Red Cross and Red Crescent Societies' (2017: 34) guide to building urban resilience highlights the importance of "mainstreaming of DRR/DM into policies, plans, regulatory frameworks and development agendas".

Yet much of this emphasis on mainstreaming resilience is centred on land use and urban development planning through formal governance instruments such as building regulations, master plans or city development plans. This is at odds with the reality of urban centres of the Global South, which are commonly characterised by a high degree of informality in labour, capital, land and entrepreneurship. The extent of informality in cities of the Global South can be seen through one of its physical manifestations – informal settlements. Thirty per cent of the world's urban population now lives in informal settlements (Ramin 2009; World Bank 2019b).

Formal planning mechanisms may also cast the informal sector as a problematic, disadvantaged consequence of labour markets, thereby failing to capture the opportunities of the informal, unregulated, dynamic micro-entrepreneurial sector for building resilience (Maloney 2004; Fraser 2018). It is becoming increasingly clear that formal planning approaches in the Global South are not able to match urban population and infrastructure growth, which "consistently outstrips even the most perspicacious planner's vision for it" (Roy 2009: 77). This suggests that formal plans and processes, as they stand, can only go so far in building resilience in urban centres across the world.

This chapter demonstrates the challenges of a disproportionate reliance on traditional and formal planning instruments for enhancing resilience and presents the case for pivoting towards planning approaches that embrace informality.

4.1 THE STATE OF PLAY: URBAN PLANNING INSTRUMENTS FOR ENHANCING RESILIENCE

Urban planning can be understood as a "decision-making process aimed at realising economic, social, cultural and environmental goals through the development of spatial visions, strategies and plans and the application of a set of policy

TABLE 4.1 The state of play: Examples of urban planning instruments for enhancing resilience

	Hazards	Exposure	Vulnerability
What is built	• Green infrastructure	• Green infrastructure • Urban redevelopment	• Emergency management • Critical services
How it is built	• Green infrastructure	• Building codes • Green infrastructure	• Enhancing accessibility and connectivity
Where it is built	• Transfer of development rights	• Zoning • Land subdivision	• Land sub-division • Buyout programmes • Transfer of development rights

principles, tools, institutional and participatory mechanisms and regulatory procedures" (UN-Habitat 2015: 2). This section outlines the state of play, providing examples of the use of urban planning tools to determine what is built, how it is built and where it is built (summarised in Table 4.1).

Among the range of planning instruments, the master plan has been salient across cities of the Global South. This provides a conceptual layout to guide future growth and development of an urban centre. The plan typically adopts a time horizon of 15 to 30 years and outlines: the locations, uses, densities and lot sizes of residential and commercial areas; the location, size, function and scale of open spaces and public areas; approaches for protecting, enhancing, managing and balancing development with biodiversity and environmental sustainability; a vision for the provision of utilities (water, electricity, gas, sewage and telecommunications infrastructure) and transport (World Bank 2019a). By contrast, city development plans are more operational, have a comparatively shorter time horizon of five years and are more focused on laying out urban development projects and financing strategies (Meshram 2006). Other types of plans include comprehensive plans (that lay out a vision for the future of a region, spanning land use, demographic shifts, culture and human development) and local development plans (that usually lay out the development vision of fairly contained spatial entities) (Goodman and Freund 1968).

Such plans have served as the entry points for mainstreaming resilience, before or after disturbances (Storbjörk and Uggla 2015). The process usually begins with risk assessment in which hazards, exposure and vulnerability are gauged for a spatial unit. This is followed by an analysis and planning phase that entails situation analysis, visioning and goal setting, analysis of land development scenarios and land use policy formulation (ADB 2016). Finally risk and resilience is integrated into the selected planning instrument by determining what is built (i.e. development control), how it is built (i.e. design control) and where it is built (i.e. location control) (Bahadur et al. 2016). We examine planning and resilience in these three spheres in the subsequent sections.

4.1.1 Development controls for resilience

Urban plans have been used to determine the buildings and infrastructure built in an urban centre. Thereby these plans determine "what is built" in a town or city to ensure that it is better able to deal with climate-induced shocks and stresses. Such development controls can enhance resilience by supporting emergency response and recovery. For example, urban planning instruments have been used to mandate the installation of flood warning systems and the development of new roads and bridges that improve capacity for evacuation (March et al. 2018). Urban plans can also include the development of emergency shelters, either through multipurpose infrastructure such that schools or sports facilities can be used flexibly in times of emergency or by designating multi-use open spaces that must be preserved for effective evacuation during emergencies (ADB 2016). Urban plans have also enabled resilience by mandating the development of critical services both after shocks have occurred and in non-emergency times; these services include hospitals and health clinics, distributed water and power supply infrastructure, law and order infrastructure, communications and mobility infrastructure (ADB 2016). These facilities help reduce the vulnerability of urban residents as they are better able to anticipate and absorb climate-induced shocks and stresses.

Urban planning instruments are increasingly mandating the development of green infrastructure providing a range of ecosystem services that include resilience to climate-related shocks and stresses (McPhearson et al. 2015). Green infrastructure reduces exposure – for instance, by providing insulation and regulating temperature – and hazards – for instance, by regulating storm water runoff and reducing flood risk downstream. Designs can range from small features to entire functional ecosystems and examples include rain gardens (planted areas in that are designed to collect and manage stormwater), stormwater green streets (similar to rain gardens but usually larger), subsurface detention systems (embedded stormwater chambers that can regulate runoff) and the delineation of green belts or functional ecosystems (designated natural areas for maintaining essential ecosystem services) (Koc et al. 2017; NYC 2019). The city of Ahmedabad in India has been severely impacted by heatwaves in the past decade, and therefore its latest city development plan emphasises the installation of green infrastructure (Mell 2018). This includes the development of green belts, urban groves, regeneration of local waterbodies and a rapid expansion of tree cover for streets. In this way, by reducing the likelihood of extreme temperature events and floods, urban plans have been used to reduce hazard risk.

Urban planning instruments have also been used to enhance resilience after disaster events have occurred, often through urban redevelopment that rezones a given area to alter density and provide infrastructure improvements after disasters (World Bank 2015a). The redevelopment plan for Christchurch, New Zealand, following a spate of devastating earthquakes in 2010–2011 spanned changes including: the construction of low-rise buildings; introducing more green space; declaring hazardous land parcels as uninhabitable; and the institution of new building standards to ensure a higher degree of structural resilience in all new construction (Gjerde 2017).

4.1.2 Design controls for resilience

As well as influencing *what* is built, urban planning instruments have been used to enhance resilience through determining *how* a town or city is built. Such design controls include the institution of building codes – regulations governing the design, construction, alteration and maintenance of infrastructure – that have been employed widely for enhancing resilience (FEMA 2019).

Building codes have been used in different ways to reduce exposure across the world. For example, the International Building Code integrates protection against multiple shocks and stresses, including: increasing resilience to hurricanes (through, for instance, mandating particular kinds of nailing patterns for roof decks or wall sheathing to enhance wind resistance); specifying ventilation and insulation standards to battle heatwaves; mandating the installation of water reuse systems and reclaimed water systems to build resilience to drought; and identifying elevation requirements to reduce exposure to flood risk (ICC 2019). One example is seen in Bangladesh, where the National Building Code aims to reduce the "climate sensitivity" of buildings by underlining that buildings must be able to withstand extreme wind, precipitation and temperatures; it also details specifications for building techniques, such as the use of confined masonry that enhances the ability of buildings to withstand shocks and stresses (Ahmed et al. 2018; Islam and Hossain 2013).

Design controls have also been employed to ensure that new developments in urban centres support green infrastructure development that helps reduce hazards and exposure. For example, codes can mandate that certain kinds of new developments must have green roofs containing live plants. Such roofs provide resilience-enhancing benefits that include absorbing excessive rainwater during extreme rainfall events, reducing ambient temperature and mitigating the risk of heat islands, increasing biodiversity and enhancing insulation (Gill et al. 2007). New York City, as part of its nodal urban plan, OneNYC, has mandated that all new buildings in the city will need to have green roofs or install photovoltaic cells (Velazquez 2019). Similarly, controls can be established for green walls that grow vegetation vertically, providing insulation and regulating ambient temperature to mitigate risk from extreme temperature events (Mustonen 2017). Design controls can also expand the coverage of blue roofs, designed to provide stormwater detention by controlling runoff while simultaneously recharging aquifers and addressing drought risk (Blick et al. 2004).

Design controls can also aim to enhance accessibility and connectivity so that "highly connected and porous urban areas facilitate better movement of people and vehicles", leading to more effective emergency response and evacuation following disturbances (Yamagata and Sharifi 2018: 18). For example, in Florida the authorities have mandated road widths to facilitate evacuation in certain areas and, therefore, reduce the vulnerability of surrounding populations (Helderop and Grubesic 2019). Finally, design controls can be instituted before disturbances or ex post through sustainable urban retrofitting or re-engineering; that is, altering the built environment in order to improve energy and water efficiency (Eames 2011).

4.1.3 Location controls for resilience

Location controls through zoning, land subdivision, relocation programmes and transfer of development rights have been used the world over to reduce exposure, vulnerability and hazards through determining *where* development takes place in a town or city.

Zoning is the most important example of location controls, permitting particular land uses in allocated zones to shape urban layout and enable different types of development (World Bank 2015b). Zoning can be critical in reducing exposure, including through incentive zoning where developers can be induced to build with greater density in areas that are relatively safe; this ensures that property in relatively safe locations is affordable, allowing urban populations to move out of hazardous land to less exposed locations (Clark 2007). In contrast, regulatory zoning can simply outlaw the development of certain land parcels to reduce the exposure of people and assets to shocks and stresses. This was precisely the approach followed by Christchurch, New Zealand, after a major earthquake in 2011, when a red zone (no construction permitted) was created within the land use plan to cover land that was highly unstable (Saunders and Becker 2015).

Closely associated with zoning are directives within urban plans for land subdivision. This refers to the process of splitting up or assembling land for development, controlling the density, configuration and layout of land parcels to regulate development in exposed areas and minimise risk (ADB 2016). Through conversion of plots of land into smaller units, urban plans can increase population density in parts of a city that are less exposed to hazards. Subdivision can also reduce vulnerability by reconfiguring urban neighbourhoods and districts to improve access and connectivity; for example, by adding or widening roads or through improvement of urban services. This has been used in a number of places, such as in Barcelona, where urban densities were altered using land subdivision (Siavash 2016).

While zoning and land subdivision for enhancing urban resilience can happen in advance of shocks and stresses, location controls can also be implemented ex post. For example, buyout programmes enable governments to purchase hazard-prone land, permitting the residents to move out of exposed land parcels and into safer territory. There are numerous examples of this sort of "adaptation through acquisition" (Moscovitz 2018). NY Rising buyout and acquisition programmes were launched in New York to address the damage caused by hurricanes Irene and Sandy as well as Tropical Storm Lee between 2011 and 2013 (Freudenberg et al. 2016). By its fifth anniversary, the programme had spent more than USD 250 million to purchase 650 affected and exposed properties to ensure "managed retreat" for enhanced resilience (GOSR 2017). A variation of the same approach exists in Vietnam, where the government provides interest-free loans to hazard-exposed, poor households on sections of the Mekong Delta so that they can relocate and purchase dwellings in safer areas (Chun 2015).

Closely associated with this approach is the transfer of development rights (Division of Local Government Services 2015). This is a technique that can be used in urban plans to protect lands that have ecological value (and therefore can help mitigate hazard risk) or are highly exposed. It works "by redirecting development that would otherwise occur on this land (the sending area) to an area planned to accommodate growth and development (the receiving area)" (Theilacker 2019: 1). Landowners receive market-determined financial compensation for not developing their land, and the development rights are sold to builders that are able to develop in other designated areas (Theilacker 2019). This system helps limit exposure and has been used widely in cities across the United States for over 100 years (Colavitti and Serra 2018).

4.2 CHALLENGES: INFORMALITY AND THE LIMITS OF URBAN PLANNING FOR ENHANCING RESILIENCE

The previous section demonstrated that urban planning instruments have employed a range of techniques to enhance resilience. However, we argue that the limitations of such approaches pose a major challenge and that they can even amplify risks and vulnerabilities in cities of the Global South. A key element of this argument is that the world's cities most vulnerable to climate change tend to be those with a high degree of informality, and formal planning instruments have limited influence in the development of large swathes of these urban centres. Informality can be understood "as settlement and activities which are not recognised by the state or do not meet state regulations"; this characterises the new normality of urban life under rapid urbanisation (Fraser 2018: 117). Informality is a foil to Euro-American planning theory that dominates urban planning regimes across the world and is characterised by qualities such as instability, indistinctness, dynamism, mobility, temporariness, recyclability and reversibility (Lutzoni 2016). Guha-Khasnobis et al. (2006) defines the informal space as beyond the purview of government mechanisms, which implies an organisational model that is devoid of rigid structure and hierarchical configuration.

In the 1950s and 1960s, discussion of informality began in the context of movement of labour into cities. Early work by economists such as Lloyd George Reynolds examined informality from the perspective of economic activity that was ad hoc, anarchic and beyond the purview of formal economic systems (Al Sayyad 2004). This discourse matured in the 1970s when major international organisations such as the International Labour Organization started to engage with these issues and employed the term "informal economy" in official research papers (Lutzoni 2016). In the 1970s and 1980s, discussions on informality were dominated by the Dualist School, which believed that informality represented traditional and pre-modern modes of lives and livelihoods that would disappear with the march of time and the progress of industrialisation (Boels 2016). From the late 1980s, this discourse has developed into an understanding of informality as a set of positive forces to be harnessed for their entrepreneurial potential and which can make a valuable contribution (Maloney 2004; Lutzoni 2016). In parallel, an understanding of informality started to expand beyond just economic activity, taking in geographic entities, spatial categories and urban planning strategies (Roy 2009). It was understood that "the formal processes of planning supply precise rules and directions for structuring the territory, while informal ones model, occupy and generate space following principles like spontaneity and self-organisation" (Lutzoni 2016: 10).

Building on this understanding of the dynamism and potential of the informal sector, in the following sections we highlight how its extent and importance in cities of the Global South challenge the presumptions around the effectiveness and equity of traditional planning-based approaches in building resilience. These challenges expose how misplaced faith in formal planning is increasingly at odds with urban environmental, economic, social and political realities. We divide these into functional and structural challenges, as summarised in Table 4.2.

TABLE 4.2 Informality and the limits of urban planning for enhancing resilience

Nature of challenge	Examples
Functional	The extent of informal areas limits the reach of formal planning The informal sector is a source of innovation and solutions that formal planning ignores
Structural	Formal plans can accentuate inequalities Formal plans privilege expert knowledge

4.2.1 Functional challenges

The scale of informality across the world directly calls into question the value of a sharp and singular focus on traditional urban planning instruments as a mechanism for achieving resilience comprehensively and sustainably. In lower-middle-income countries, both formal and informal activity is responsible for urban expansion of towns and cities (Roy 2010). However, with one in three urban residents across the world living in informal settlements, the influence of formal planning is, clearly, limited. This is particularly important as these contexts are the reality for many of the world's poorest and most climate-vulnerable citizens. Certain geographical contexts that are considered highly vulnerable are precisely those that have the highest degree of informality. For instance, Guyana, Guatemala, the Philippines and Bangladesh all figure in the top ten of the World Risk Index and are marked by 33%, 35%, 38% and 55%, respectively, of their urban populations living in informal settlements (Day et al. 2019; World Bank 2019b). This picture is stark at the regional level too, as 56% of the entire urban population of sub-Saharan Africa lives in informal settlements and the figures for South Asia and Latin America are 31% and 21%, respectively (UN-Habitat 2016). Thus, the lives and livelihoods of many people living in urban areas of the Global South are not governed through formal urban planning instruments. Moreover, marginal populations in many of the world's cities cannot comply with regulation and planning and have no option but to reside on highly risk-prone sites such as steep slopes and floodplains (Yamagata and Sharifi 2018).

Despite the evidence of limited reach of formal planning instruments and ubiquity of informal settlements across the

Global South, planning regimes often ignore informality or attempt to bring these areas into the formal system. In his review of planning systems in China, Wang (2015) notes that the literature on planning in China centres around an exploration of large, state-led urban development initiatives in major cities while far less time is spent on examining the informal or even illegal planning and development practices that have typified urbanisation in the country.

While the fact that formal planning instruments simply do not reach large numbers of people is one functional challenge; another is that excluding or marginalising this sector in planning can be self-defeating as the informal sector can play a valuable role in enhancing resilience and urban development. Informal practices and knowledge have been seen to unlock bottom-up innovation in cities, and this needs to be harnessed and expanded instead of being excluded and suppressed by formal processes (Roy 2010). Song (2016) illustrates this with a study in Solo, Indonesia, where a new Bus Rapid Transport (BRT) corridor was developed with the help of a team of international technical consultants. In seeking to develop this modern and ostensibly more efficient public transport system, the plans jeopardised the existing, informal system of *angkots* (minivans or modified pickup trucks with added seats). There was very poor uptake of the BRT, and an investigation revealed that the reason for this was the marginalising of *angkots* and lack of "last mile" connectivity from people's homes and offices to BRT stations (especially in some of the peripheral areas of the city). Moreover, over half of all *angkot* riders were women and most earned a below-average income. The vital role played by *angkots* in extending transport facilities to the poor reveals the need for diverse transport systems to work together (Song 2016).

There is a growing understanding that supporting the informal sector can deliver novel and effective solutions. Informal processes of innovation that are bottom-up, indigenous and suited to local cultural norms, inexpensive and frugal, developed through subjective processes that rely on the innovator's intuition, can be effective mechanisms for enhancing resilience (Bahadur and Doczi 2016). Heisel (2016: 30) draws on this thinking by presenting a new understanding of formal city planning within a framework of locally developed knowledge, traditions, economic systems and social and cultural networks. He presents a powerful argument for embracing the informal, stating that this could support the development of a bottom-up and contextually relevant vision of urban resilience and tap into the skills, labour and capital of the informal sector that has previously been employed illegally (Heisel 2016). We contend, therefore, that an exclusive reliance on formal planning can reduce functional efficiencies in resilience building by marginalising informal processes, when in fact accepting and encouraging these processes carries the potential to deliver more effective and equitable urban resilience.

4.2.2 Structural challenges

Apart from the functional challenges, urban planning instruments also present a number of deeper structural and endemic challenges. One such issue is that formal planning can deepen inequalities. Urban redevelopment, regeneration and upgradation initiatives tend to focus on physical capital and infrastructure, with less attention to issues of livelihoods, rights and political participation of the urban poor (Song 2016). Research in the informal settlements in Korail, Dhaka, found that relocation initiated by the government is not considered favourably by residents as this disrupts livelihoods and social networks that have taken years to consolidate (Jabeen et al. 2010). Formal relocation efforts frequently misunderstand the priorities of those living in informal areas, where access to livelihood and social ties tend to trump upgraded infrastructure in peripheral urban locations.

Attempts to formalise the informal through processes such as land titling can inadvertently reinforce gender inequalities and trigger conflict at the household level. This is because most initiatives that give titles to those occupying informal settlements consider the head of the household (usually an elder male) to be the rightful recipient of the formal title to the land. While informal systems are not necessarily essentially more harmonious and equitable, formal property systems can be embedded with patriarchal and class power, accentuating gender inequities (Roy 2005).

Formal planning processes can also deepen inequality by inducing the displacement of some of the most vulnerable urban residents. For instance, in 2004, the Angolan state issued a law that turned all unregistered land into state property as part of a process to legalise property rights. This in turn led to many of the poorest urban residents being confronted with forced evictions and demolitions (Koster and Nuijten 2016). There are repeated instances globally of how formal planning processes take an arbitrary approach to declaring certain lands as informal as a precursor to demolitions and displacement. Studies across Indian cities note how the state often determines what is informal and what is not, thus allowing the sprawling of elite farmhouses (that contravene a number of laws) on the edges of Delhi to function while slum colonies are frequently demolished (Roy 2009). In this way, formal planning processes have been seen to amplify socio-economic inequalities in cities.

Formal planning processes commonly prioritise the knowledge of experts and technocrats while sidelining others (Meerow and Woodruff 2020). Reliance on technical knowledge and formal, planned approaches can lead to the prioritising of structural solutions (e.g. dams and levees) to urban risk. This, in turn, can lead to a false sense of security, as these approaches accommodate risks within certain ranges and may not be able withstand the sort of historically unprecedented weather and climate events that are becoming more common (Yamagata and Sharifi 2018). One of the key reasons for overlooking the value of the informal *angkot* system in Solo, Indonesia, was the reliance on scientific and expert knowledge from teams of technical consultants without taking account of the lived experience of the ordinary citizen. This can marginalise citizens from decision-making processes and lead to failures as the formal system imposes externally defined solutions or formalises informal systems without acknowledging local, contextual factors. Roy (2005) links the tendency of formal planning regimes to privilege certain kinds of knowledge to the inherent Eurocentricity of the discipline of urban planning. Much of the urban growth of the 21st century is taking place in developing countries, but the salient theories of how cities function that planners employ remain rooted in the contexts of developed countries. As Song (2016: 377) notes,

> [r]ationalistic, scientific, intellectual investigation and analysis constitute only one route among several to solving social problems; ordinary knowledge invoking common sense, casual empiricism or thoughtful speculation and analysis, along with social organisation and interaction incorporating various habits, traditions, customs or routines, are just as – if not more – vital.

The dominant reality of urban areas is one of flexible spatial arrangements, spontaneity and informal expansion, suggesting there is a need for planning to employ knowledge in ways that are more tactical and adaptable, rather than being strategic and rigid (Lutzoni 2016). The employment of different kinds of knowledge and the engagement of diverse constituencies in planning enhances possibilities to surface different types of solutions that are better able deal with a variety of expected and uncertain shocks and stresses induced by a changing climate. The American Planning Association recognises climate change as one of the most important planning challenges of the 21st century and underlines the importance of planners engaging with communities to find solutions (Meerow and Woodruff 2020). Even though most urban planning processes solicit the views of citizens through public consultation exercises, these are mostly shallow, technical attempts to tick the "participation" box. Commenting on this, Cohen et al. (2018: 16) note that "participation cannot just be a slogan, it must be designed as an integral part of each of the steps that make up an urban practice oriented towards risk management".

4.3 EMBRACING INFORMALITY FOR ENHANCING RESILIENCE

The combination of challenges outlined in Section 4.2 demonstrates that even though formal approaches are used in good faith as instruments to bring resilience considerations into spatial planning, formal plans frequently fail to deliver benefits to significant proportions of the urban population. We argue that functional and structural challenges with traditional urban planning can be overcome by recognising informality, namely how informality can be incorporated and embraced in planning processes for enhancing resilience as opposed to being either bypassed or co-opted into an existing vision of the formal. Doing so requires applying new kinds of knowledge in planning processes and implementing novel modes of planning that involve new constituents in decision-making (Meerow and Woodruff 2020). This represents a new approach for urban development that harnesses informal practices and informal economic activity to build resilience.

4.3.1 Embracing informal knowledge in planning for resilience

One of the key structural challenges with urban planning is the reliance of urban planning systems on expert knowledge and the fact that the perspectives of the ordinary citizen, especially those deemed to be informal, are frequently overlooked. There is a growing number of examples from around the world demonstrating the mechanisms and benefits of including the perspectives of those occupying informal spaces.

The city of Antofagasta, Chile, is highly exposed to flooding and earthquakes (Bozkurt et al. 2016). At the same time, as it is a port city and has a booming economy, it has a growing population that stretches ecological boundaries and leads to heightened climate risk. Given that the urban poor population nearly doubled between 2011 and 2018 across Chile, the city administration realised that this problem could not be solved without the inclusion of informal knowledge into urban planning processes for resilience (De la Jara 2018).

Therefore, the city administration partnered with an urban planning NGO to design and execute a highly inclusive approach to planning for resilience that was predicated on deep participation with marginalised urban populations (Rosenzweig et al. 2018). As part of this, the stakeholders leading the initiative decided to host large, open-air street meals (known as Malones Urbanos) that built social capital but, more importantly, provided a platform for collaborative problem-solving on risk and resilience. Across a number of such events, this initiative helped develop a comprehensive understanding of diverse topics – from placemaking to land reclamation and from climate adaptation to waste management – in a way that was engaging, participatory and yielded results that could directly inform policy and planning processes for enhancing resilience. As well as these communal meals, large, open-air events (known as Okuplazas) were held in marginalised parts of the city (Ciudad Emergente 2020). These were used to develop diagnostic tools and indicators, socio-environmental maps, "idea trees" for tracking local needs and conservation priorities. These specific tasks of collating local knowledge were interspersed with cultural performances to promote engagement and enthusiasm.

In this way, tools and approaches steeped in an understanding of local culture were employed to solicit meaningful participation of those at risk, thus embodying an important principle of strong climate change planning (Meerow and Woodruff 2020). This not only led to the collection of accurate data with the potential to influence resilience plans to benefit different sections of the city's population, but also helped secure the buy-in of communities at risk, which is essential for the success of any resilience planning exercise. This aligns very closely with Cohen et al.'s (2018: 17) observation that "urban resilience cannot be an objective of policies developed outside the community, rather it must be the product of local processes".

While participatory approaches based on local customs are one innovative mechanism through which urban plans are including informal knowledge to enhance resilience, the inclusion of data from self-enumeration exercises provide another (Beukes 2015). Since official surveys and censuses tend to cover areas that have legal status, information on informal areas is often not available to municipal authorities. The de facto invisibility of these areas makes it more likely that authorities can undertake demolition, eviction, redevelopment or upgrading initiatives that are at odds with local realities, as they are based on inaccurate information. These outcomes put the residents of informal areas at risk. As a result, self-enumeration exercises – where informal communities collect, standardise, analyse and present information about the areas in which they live and work – have become increasingly popular. Most of these surveys capture settlement profiles (community assets, nature of poverty, history of settlement, infrastructure and services); household census (household size, income, assets, access to basic services, and tenure security); and "mapping" (geographic information on settlement boundaries, structures and basic facilities such as toilets, schools and surface water). At last count, over 7,000 settlements had undertaken such exercises across 15 countries (Beukes 2014).

There is a growing number of examples of the how this data is influencing planning decisions with knock-on effects for resilience. For instance, in a settlement in Accra, Ghana, a self-enumeration exercise revealed that there were far more people than expected living in an informal settlement that was marked for eviction, and this knowledge led to the government pausing and rethinking their plans (Provost 2012). In contrast, a survey of a settlement in Cape Town, South Africa, showed far lower numbers than expected, suggesting that upgrading would be easier and quicker than relocation. There are number of other instances where governments have delivered urban services (e.g. sanitation) to informal populations based on evidence from such exercises (Makau et al. 2012). Interestingly, methods resembling this proved valuable in Dharavi, Mumbai (one of the largest slums in Asia) during the COVID-19 crisis. Here, teams of local volunteers familiar with the neighbourhood went house to house testing residents and gathering information for the government. The tactic proved indispensable in containing the virus in this vast informal settlement.

BOX 4.1 Street vendor relocation in Surakarta, Indonesia

History rarely offers natural experiments, but one such unique case comes from Surakarta, Indonesia, where there was a sustained programme to relocate street vendors (the backbone of the informal economy) from the city centre to an area specially designated for them in another part of the city. The relocation happened in two waves. In 2005, the urban planners and city officials led by the mayor undertook extensive and iterative consultations with the street vendors and their representatives to understand the reasons for their hesitancy in moving to the newly allocated space. The street vendors' suggestions were incorporated and requests furnished to the extent possible. This led to successful relocation and a flourishing trade for these vendors in the new space. In 2012, a second wave of relocation took place, but due to transitions in the city's political and bureaucratic establishment, the process was top-down and without earnest engagement with the street vendors. This resulted in high rates of market abandonment and stall vacancies due to numerous inadequacies in the relocation areas. This ultimately led to street vendors returning to the areas they previously occupied.

Source: Adapted from Song (2016)

These examples demonstrate mechanisms through which informal knowledge can be appropriated into formal planning initiatives. Crucially, they prove that the imperative for this goes beyond the esoteric, and the functional success of various urban plans and policies is predicated on the use of such methods. In Antofagasta, Chile, deep and sustained community engagement was the only way to comprehensively build resilience in this rapidly expanding city with large numbers of poor and marginalised citizens. Self-enumeration exercises have demonstrated time and time again that local communities are best positioned to provide vital information about their neighbourhoods; and by extension, this permits a more precise understanding of exposure and vulnerability. This is why Dobson and colleagues (2015) argue that the lack of data on the nature of urban poverty and vulnerability in developing countries is one key factor that impedes resilience. Self-enumeration processes could thereby fill a critical need.

4.3.2 Embracing informal actors in planning for enhancing resilience

Apart from the appropriation of new kinds of knowledge that allows urban planning instruments to better reflect the needs and priorities of those deemed informal, we develop an evidence base to contend that the engagement of non-traditional actors in developing and executing urban plans can deliver rich dividends for resilience-building objectives.

One example is barefoot planners and architects. This concept is derived from the "barefoot doctors" in China, who received rudimentary training in order to meet the ballooning health needs of China's large and expanding population. Barefoot doctors "lived in the community they served, focused on prevention rather than cures while combining western and traditional medicines to educate people and provide basic treatment" (Weiyuan 2008: 909). Similarly, there is a growing call for the mainstreaming of barefoot planners and architects. These semi-professionals with a more intimate knowledge of local geographies and customs are better able to comprehend rapidly expanding informal areas in cities of the Global South. Examples from countries such as Swaziland, China and Thailand are demonstrating how these semi-formal, community-based professionals need to be embraced as valuable tools for effective planning for resilience.

For instance, as part of a project to upgrade informal settlements in Mbabane, Swaziland, a group of local volunteers were given basic training and asked to play the role of neighbourhood upgrading facilitators. These individuals were then tasked to first develop a detailed map and census of the local area and then undertake detailed discussions with residents to propose improvements that reflected local priorities and aligned with local values (Martin and Mathema 2009). Their designs and inputs were then translated into computer-aided design formats that could be used in formal urban planning. Similarly, the Asian Coalition for Community Action Programme, that has supported upgrading initiatives by forging partnerships between communities and governments in 107 cities across 15 countries, has made wide use of barefoot planners and architects. Here, individuals from within communities that are to be upgraded undertake mapping, surveying, planning and designing functions with some support from professionals, leading to solutions that are a better fit for local residents (Archer et al. 2012).

This approach is also illustrated in slum development and risk reduction initiatives in Cape Town, a context with a high degree of mistrust between the authorities and residents of informal settlements. Ziervogel (2019) documents how, in order to enhance resilience to flooding, the government planned on shifting one particular high-risk settlement to large, elevated gravel platforms. In order to understand community needs and to build an atmosphere of trust, they recognised the importance of employing a respected local leader as an intermediary. This local leader helped translate the government's plans for the community, communicated community demands to the government and, over time, aligned the interests of different groups to build a shared discourse of change to ensure this shift. Through his ability to read

> both the needs of the settlement residents as well as what the City could support, he was able to shift the trajectory of services in the settlement towards what residents wanted, rather than waiting for the City to decide on their priorities.
> (Ziervogel 2019: 499)

A slightly different vision of barefoot planning is evident in the Chinese city of Xinxiang, where the government expropriated land from peri-urban farming communities for the development of an industrial park. As large parts of the expropriated land remained vacant due to lack of investment in the planned industrial park, a leader from the dispossessed farming communities developed informal arrangements with local governments that allowed erstwhile farmers to return to farming (Wang 2015). Due to this flexible land use solution worked out by this barefoot planner, the farmers were able to return to their livelihoods and the local government was able to meet its food production quota (issued by Beijing) without any official changes or rollbacks in zoning.

These examples all demonstrate how the involvement of non-traditional actors in the development, execution and mediation of urban plans can help bridge the divide between formal planning regimes and informal spaces. Professional planners should learn from barefoot planners, who employ contextually relevant negotiation and consensus-building tactics to solve pernicious problems of planning for resilience (Wang 2015). While the impacts on resilience are fairly direct in the Cape Town example, in the Mbabane and the Asian Coalition for Community Action Programme examples, the inclusion of local intermediaries in planning processes meant that planners acquired a much more precise understanding of exposure and vulnerability in these otherwise invisible areas. Also, by acceding space to the barefoot planner in Xinxiang, the government was able to find an endogenous solution for making local livelihoods more resilient and strengthening the local food system.

During COVID-19 too, innovative partnerships between formal and informal actors have emerged. For instance, in parts of Mumbai, the Municipal Corporation partnered, for the first time, with local medical practitioners working within informal settlements. The agency provided these practitioners with training on managing COVID-19 and developed systems to ensure that infected patients were rapidly referred to government hospitals. In this way, government agencies were able to draw on the local knowledge, networks and relationships of trust that these actors enjoyed to reach those living outside formal systems and mount an effective response to the pandemic (ANI 2020).

4.3.3 Embracing informal practices in planning for enhancing resilience

Along with new kinds of knowledge and actors, it is vital to embrace informal practices. These include ways in which the informal sector determines how urban centres expand and the nature, function and form of informal infrastructure. A growing number of examples show how appropriating and shaping these practices can help deliver rich dividends.

Across the developing world, towns and cities are comprised of a substantial proportion of non-engineered buildings built by informal actors (with minimal formal training and no certification) and with no adherence to any monitoring and inspection protocols during construction (Shaw et al. 2005). The scale of these spontaneously constructed buildings, that do not adhere to building codes or standards, is significant. By some estimates, in urban areas of large lower-middle-income countries such as India, one in every four buildings is non-engineered or informal, and a large proportion of these buildings are situated on hazard-prone land (Shaw et al. 2005). Recognising that a singular focus on developing and promulgating better building codes is likely to have a limited impact on enhancing urban resilience, a number of organisations are starting to shape informal building practices by training and building the capacity of casual masons and builders responsible for the construction of non-engineered buildings.

After the devastating earthquake of 2001 in the Indian state of Gujarat claimed over 30,000 lives and led to the collapse of over 300,000 buildings, SEEDS (Sustainable Environment and Ecological Development Society) India built the capacity of informal builders and masons to use more resilient building techniques. Shaw and Izumi (2014) describe how, in collaboration with architects and civil engineers, SEEDS India trained masons in seismic-resistant building materials and approaches. To incentivise training and, in turn, ensure sustainability, the government certified trained masons, offered them identification cards and laid the foundation for a Mason's Association into which trained and certified masons were enrolled. This created an easily accessible database of certified masons, recognised masonry as a skilled profession and created links with local construction companies. Similarly, in Medellín, Colombia, Build Change (an organisation focused on promoting resilient building techniques) partnered with the National Learning Service of the Ministry of Labour to provide technical training on seismic evaluations and resilience-enhancing retrofitting approaches through a theory-based module followed by on-the-job training to builders and masons (Build Change 2018). The organisation also trained homeowners in the principles of disaster-resistant house construction so that they would have the knowledge and skills needed to supervise disaster-resilient construction.

In addition to shaping and improving informal building practices, another strategy for enhancing resilience that goes beyond the purview of formal planning involves embracing new modes of urban redevelopment and regeneration. An example of one such approach is the in situ upgrading of informal settlements. Here, the housing rights of the residents of informal settlements are recognised, and they are provided many of the urban services, infrastructure improvements and social schemes that are available to more formal parts of the city in a way that entails minimum disruption to their lives and livelihoods (Del Mistro and Hensher 2009). A famous example is the Favela Bairro programme in Rio De Janeiro, Brazil, where instead of evicting and trying to relocate citizens on the urban periphery, as had been the norm, the city, in partnership with international organisations, engaged local communities in bringing civic and social services to the area (Atuesta and Soares 2018). This resulted in reduced crime, improved well-being, less inequality and, concomitantly, reduced exposure, vulnerability and hazard risk.

Another version is found in the Orangi Settlement in Karachi, Pakistan. Here a ground-up initiative led by civil society actors and local residents to upgrade basic infrastructure, benefiting close to 100,000 informal households, was recognised, adopted and scaled up by the authorities (Hasan 2019). Similarly, in situ upgrading has become a dominant approach for managing informality in Afghanistan (see Box 4.2).

BOX 4.2 In situ upgrading on informal settlements for enhancing resilience in Afghanistan

Afghanistan suffers from multiple, intersecting risks of climate change, disasters, migration and conflict; it will be impacted by increasing temperatures, extreme precipitation and concomitant flooding in the short term but expanding desertification and water scarcity over the long term (Savage et al. 2009). Moreover, the country's urban population is highly vulnerable as almost two-thirds of all residents of towns and cities live in informal settlements (UN-Habitat 2020). This is partly because the country is seeing very rapid urbanisation: the share of the country's population living in urban centres jumped from 20% in 2002 to 33% in 2015, and this will increase to over 50% by 2060 (French et al. 2019).

Until the end of the first decade of the 21st century, demolition, eviction and relocation into government housing was the main response to dealing with informal settlements across much of country. At this point, this strategy was acknowledged as having failed, as the government was able to provide an average of only 2,000 housing units a year – not nearly enough to cater to the needs of those living in rapidly expanding informal settlements. Instead, the government, with international partners, launched a nationwide participatory in situ slum upgrading policy "to strengthen urban communities' resilience and capacities in the face of limited services, conflict and underdevelopment" (French et al. 2019: 214).

Through engagement with community development councils – officially recognised local networks for development planning and implementation based on indigenous forms of social association – the policy supported the development of community action plans, participatory design of resilience-enhancing action and jointly implemented upgrading initiatives. The approach hinged on contributions from the communities in cash or in kind, amounting to between 25% and 35% of project costs. As of 2016, over 1.5 million Afghans in over 170,000 households have benefited from this upgrading, which has included the upgrading of sewer systems, paving of roads and the provision of basic services for improving well-being, reducing exposure and increasing adaptive capacity (Amiri and Lukumwena 2018; Khan 2014).

These examples all demonstrate how accepting, incorporating and embracing informal practices such as building construction, settlement and livelihood activities has helped authorities pivot away from existing approaches that were ineffective and inequitable. Instead, acknowledging the vital role of self-enumerators, barefoot planners and informal builders, recognising the rights of slum dwellers and mainstreaming local efforts of infrastructure development have delivered contextually relevant and culturally acceptable models for enhancing, at scale, the resilience of the most vulnerable urban citizens.

4.4 CONCLUSION

This chapter began with an outline of the ways in which global framings for urban resilience have employed traditional and formal urban planning instruments as a key entry point for mainstreaming. The chapter then explored the current state of play, describing how urban planning instruments have employed development, design and location controls to reduce vulnerability, exposure and hazards. Following this, the chapter argued that the high degree of informality across the cities that are most at risk of climate change renders a sharp focus on traditional, formal planning process less effective. Formal processes rely on expert knowledge, fail to leverage the innovation potential of the informal sector and accentuate inequalities. Finally, the chapter drew on a range of global examples to demonstrate how embracing informal knowledge, actors and practices can help overcome these challenges. This analysis permits the distillation of a few key insights (summarised in Table 4.3).

First, this chapter has not argued that urban planning instruments have no value as a mechanism for enhancing resilience. We showed how these have been used to reduce vulnerability, exposure and hazards across the world. Instead, the chapter made a case for better recognising the realities, rights and requirements of those deemed informal. Evidence indicates that without this acknowledgement, urban plans will not meet their objectives and those in informal areas will continue to be at risk. This apart, the chapter also underlined that despite the best-laid plans, the development of a city will be influenced by informal actors and practices; therefore, the global community of practice on urban resilience must embrace and strengthen these actors and practices.

Second, the chapter demonstrated that the relationship between the formal and informal is dynamic and multifaceted. There are at least two clear directions of travel. On one hand, it is possible to bring the informal into formal plans and processes by, for instance, using results from self-enumeration exercises in planning and establishing participatory planning mechanisms. On the other hand, it is possible to bring formal services and networks to informal areas through, for instance, the in situ participatory development of informal settlements, such as in the examples provided from Afghanistan, Brazil and Pakistan. These examples demonstrated that there is no one-size-fits-all solution; rather, how the informal and formal combine is dependent on local contextual realities, objectives of specific initiatives and the nature of actors involved.

Third, we made the case that integrating informal practices has a direct impact on resilience. For instance, the use of barefoot planners and architects in South Africa and Swaziland allows planners to get a better understanding of exposure and vulnerability at the local level. The provision of basic services in the Favela Bairro programme, the Orangi settlement and across Afghanistan has helped reduce vulnerability. Even though certain initiatives described in the chapter were not explicitly focused on reducing climate risk, they offer replicable principles that should be taken on board by resilience planners. For instance, for the street vendors in Surakarta, Indonesia, the resilience impact of relocation was unclear, but the case demonstrates the pivotal nature of community participation and the importance of employing different kinds of knowledge in decision-making processes.

TABLE 4.3 Overcoming challenges with formal planning by embracing informality

Challenges		Informal knowledge	Informal actors	Informal practices
Functional	Limits to the influence of formal planning	• Meaningful community engagement to determine adaptation solutions		• Training informal builders and masons ensures resilient construction
	Innovation potential of the informal sector		• Barefoot planners and architects find locally relevant solutions	
Structural	Formal plans can increase inequalities		• Neighbourhood slum upgrading facilitators understand trade-offs	• Participatory in situ slum redevelopment leads to more equitable outcomes
	Formal plans privilege expert knowledge	• Self-enumeration exercises by residents of informal settlements ensures the inclusion of bottom-up information in planning		

In essence, this chapter has demonstrated how an exclusive focus on mainstreaming resilience in traditional and formal urban planning instruments can perpetuate vulnerability and risk. It then provided actionable and replicable pathways for pivoting towards a new reality of urban resilience planning, one where the formal and informal co-exist in a state of positive synergy to help ensure that vulnerable communities are better able to flourish despite the demands of a changing climate.

REFERENCES

ADB. (2016). *Reducing Disaster Risk by Managing Urban Land Use.* Manila: Asian Development Bank.

Ahmed, I., Gajendran, T., Brewer, G., Maund, K., von Meding, J., Kissa, G., Kabir, H., Faruk, M., Shrestha, H. D. and Sitaula, N. (2018). *Understanding the Opportunities and Challenges of Compliance to Safe Building Codes for Disaster Resilience in South Asia: The Cases of Nepal and Bangladesh.* Kobe: Asia-Pacific Network for Global Change Research.

Al Sayyad, N. (2004). Urban informality as a "new" way of life. In A. Roy and N. AlSayyad (Eds), *Urban Informality: Transnational Perspectives from the Middle East, Latin America, and South Asia* (pp. 7–30). Lanham, MD: Lexington Books.

Amiri, B. A. and Lukumwena, N. (2018). An overview of informal settlement upgrading strategies in Kabul City and the need for an integrated multi-sector upgrading model. *Current Urban Studies*, *6*(3), 348–365.

ANI. (2020, 13 July). *Locals, doctors credit joint efforts for Dharavi getting praise from World Health Organisation for tackling COVID-19.* NDTV News. Retrieved 23 November, 2020, from https://swachhindia.ndtv.com/locals-doctors-credit-joint-efforts-for-dharavi-getting-praise-from-world-health-organisation-for-tackling-covid-19-46855/

Archer, D., Luansang, C. and Boonmahathanakorn, S. (2012). Facilitating community mapping and planning for citywide upgrading: The role of community architects. *Environment and Urbanization*, *24*(1), 115–129.

Atuesta, L. H. and Soares, Y. (2018). Urban upgrading in Rio de Janeiro: Evidence from the Favela-Bairro programme. *Urban Studies*, *55*(1), 53–70.

Bahadur, A. and Doczi, J. (2016). *Unlocking Resilience through Autonomous Innovation.* London: Overseas Development Institute.

Bahadur, A., Tanner, T. and Pichon, F. (2016). *Enhancing Urban Climate Change Resilience: Seven Entry Points for Action*. Manila: Asian Development Bank.

Beukes, A. (2014). *Know Your City: Community Profiling of Informal Settlements*. London: International Institute for Environment and Development.

Beukes, A. (2015). *Making the Invisible Visible: Generating Data on "Slums" at Local, City and Global Scales*. London: International Institute for Environment and Development.

Blick, S. A., Kelly, F. and Skupien, J. J. (2004). *New Jersey Stormwater Best Management Practices Manual*. New Jersey: Department of Environmental Protection.

Boels, D. (2016). *The Informal Economy: Seasonal Work, Street Selling and Sex Work*. Cham: Springer.

Bozkurt, D., Rondanelli, R., Garreaud, R. and Arriagada, A. (2016). Impact of warmer eastern tropical Pacific SST on the March 2015 Atacama floods. *Monthly Weather Review*, *144*(11), 4441–4460.

Build Change. (2018). *Success in Haiti*. Retrieved 23 November, 2020, from https://buildchange.org/app/uploads/2018/08/Haiti-8-Years-of-Success–2018.pdf

Chun, J. M. (2015). *Planned Relocations in the Mekong Delta, Vietnam: A Successful Model for Climate Change Adaptation, a Cautionary Tale, or Both?* Washington DC: Brookings Institution.

Ciudad Emergente. (2020). *Okuplaza*. Retrieved 23 November, 2020, from https://ciudademergente.org/build/tactics/okuplaza2

Clark, J. S. (2007). Rocking the suburbs: Incentive zoning as a tool to eliminate sprawl. *Brigham Young University Journal of Public Law*, *22*(1), Article 6. https://digitalcommons.law.byu.edu/jpl/vol22/iss1/6

Cohen, M., Carrizosa, M., Gutman, M., Leite, F., López García, D., Nesprias, J., Orr, B., Simet, L. and Versace, I. (2018). *Facing Risk. New Urban Resilience Practices in Latin America*. Caracas: CAF.

Colavitti, A. M. and Serra, S. (2018). The transfer of development rights as a tool for the urban growth containment: A comparison between the United States and Italy. *Papers in Regional Science*, *97*(4), 1247–1265.

Cote, M. and King, R. (2017). *Urban Resilience Planning for Secondary African Cities: A Case Study of Mozambique*. Washington DC: United States Agency for International Development.

Day, S. J., Forster, T., Himmelsbach, J., Korte, L., Mucke, P., Radtke, K., Thielborger, P. and Weller, D. (2019). *The World Risk Report*. Bochum: Institute for International Law of Peace and Armed Conflict.

De La Jara, A. (2018, 26 December). *Slums on the rise in Chile*. Reuters. Retrieved 23 November, 2020, from https://www.reuters.com/article/us-chile-migrants/slums-on-the-rise-in-chile-idUSKCN1OP1DJ

Del Mistro, R. and Hensher, D. A. (2009). Upgrading informal settlements in South Africa: Policy, rhetoric and what residents really value. *Housing Studies*, *24*(3), 333–354.

Dobson, S., Nyamweru, H. and Dodman, D. (2015). Local and participatory approaches to building resilience in informal settlements in Uganda. *Environment and Urbanization*, *27*(2), 605–620.

Division of Local Government Services. (2015). *Transfer of Development Rights*. New York: Department of State.

Eames, M. (2011). *Developing Urban Retrofit Scenarios: An Outline Framework for Scenario Foresight and Appraisal.* Retrofit 2050 Working Paper. Cardiff: Cardiff University.

Federal Emergency Management Authority. (2019). *Building Codes*. Retrieved 23 November, 2020, from https://www.fema.gov/building-codes

Fraser, A. (2018). Informality in the New Urban Agenda: From the aspirational policies of integration to a politics of constructive engagement. *Planning Theory & Practice*, *19*(1), 117–137.

French, M., Popal, A., Rahimi, H., Popuri, S. and Turkstra, J. (2019). Institutionalizing participatory slum upgrading: A case study of urban co-production from Afghanistan, 2002–2016. *Environment and Urbanization*, *31*(1), 209–230.

Freudenberg, R., Calvin, E., Tolkoff, L. and Brawley, D. (2016). *Buy-In for Buyouts: The Case for Managed Retreat from Flood Zones*. Cambridge MA: Lincoln Institute of Land Policy.

Gill, S. E., Handley, J. F., Ennos, A. R. and Pauleit, S. (2007). Adapting cities for climate change: The role of the green infrastructure. *Built Environment*, *33*(1), 115–133.

Gjerde, M. (2017). Building back better: Learning from the Christchurch Rebuild. *Procedia Engineering*, *198*, 530–540.

Goodman, W. I. and Freund, E. C. (Eds). (1968). *Principles and Practice of Urban Planning*. Washington DC: International City Managers Association.

Governor's Office of Storm Recovery. (2017). *Five Years Later: A Retrospective.* New York: Governor's Office of Storm Recovery.

Guha-Khasnobis, B., Kanbur, R. and Ostrom, E. (2006). Beyond formality and informality. In B. Guha-Khasnobis, R. Kanbur and E. Ostrom (Eds), *Linking the Formal and Informal Economy: Concepts and Policies* (pp. 75–92). Oxford: Oxford University Press.

Hasan, A. (2019). Orangi Pilot Project. In A. Orum (Editor-in-Chief), *The Wiley Blackwell Encyclopedia of Urban and Regional Studies*. https://doi.org/10.1002/9781118568446.eurs0227

Heisel, F. (2016). Introduction: Informality in emerging territories. In H. Basel and B. Woldeyessus (Eds), *Lessons of Informality: Architecture and Urban Planning for Emerging Territories – Concepts from Ethiopia* (pp. 14–33). Basel: Birkhäuser.

Helderop, E. and Grubesic, T. H. (2019). Flood evacuation and rescue: The identification of critical road segments using whole-landscape features. *Transportation Research Interdisciplinary Perspectives*, *3*, Article 100022.

IFRC. (2017). *Building Urban Resilience: A Guide for Red Cross and Red Crescent Engagement and Contribution*. Geneva: International Federation of the Red Cross and Red Crescent Societies.

International Code Council. (2019). *Resilience Contributions of the International Building Code*. Washington DC: International Code Council.

Islam, S. and Hossain, I. (2013). *Bangladesh National Building Code (BNBC): Current Practice and Relevance to Future Climate Change Aspect CCTC 2013*. Paper Number #1569697747.

Jabeen, H., Johnson, C. and Allen, A. (2010). Built-in resilience: Learning from grassroots coping strategies for climate variability. *Environment and Urbanization*, *22*(2), 415–431.

Khan, F. (2014). Adaptation vs. development: Basic services for building resilience. *Development in Practice*, *24*(4), 559–578.

Koc, C. B., Osmond, P. and Peters, A. (2017). Towards a comprehensive green infrastructure typology: A systematic review of approaches, methods and typologies. *Urban Ecosystems*, *20*(1), 15–35.

Koster, M. and Nuijten, M. (2016). Coproducing urban space: Rethinking the formal/informal dichotomy. *Singapore Journal of Tropical Geography*, *37*(3), 282–294.

Lutzoni, L. (2016). In-formalised urban space design: Rethinking the relationship between formal and informal. *City, Territory and Architecture*, *3*(1), 1–14.

Makau, J., Dobson, S. and Samia, E. (2012). The five-city enumeration: The role of participatory enumerations in developing community capacity and partnerships with government in Uganda. *Environment and Urbanization*, *24*(1), 31–46.

Maloney, W. F. (2004). Informality revisited. *World Development*, *32*(7), 1159–1178.

March, A., Nogueira de Moraes, L., Riddell, G., Stanley, J., van Delden, H., Beilin, R., Dovers, S. and Maier, H. (2018). *Practical and Theoretical Issues: Integrating Urban Planning and Emergency Management – First Report for the Integrated Urban Planning for Natural Hazard Mitigation Project*. Australia: Bushfire and Natural Hazards CRC.

Martin, R. and Mathema, A. (2009). *Development Poverty and Politics: Putting Communities in the Driver's Seat*. New York: Routledge.

McPhearson, T., Andersson, E., Elmqvist, T. and Frantzeskaki, N. (2015). Resilience of and through urban ecosystem services. *Ecosystem Services*, *12*, 152–156.

Meerow, S. and Woodruff, S. C. (2020). Seven principles of strong climate change planning. *Journal of the American Planning Association*, *86*(1), 39–46.

Mell, I. C. (2018). Greening Ahmedabad: Creating a resilient Indian city using a green infrastructure approach to investment. *Landscape Research*, *43*(3), 289–314.

Meshram, D. S. (2006). Interface between city development plans and master plans. *ITPI Journal*, *3*(2), 1–9.

Moscovitz, A. N. (2018). *Adaptation through acquisition: Planning for home buyout and acquisition in the New York Region* [PhD thesis]. Columbia University.

Mustonen, T. (2017). *What are Green Walls – the definition, benefits, design and greenery*. NAAVA. Retrieved 23 November, 2020, from https://www.naava.io/editorial/what-are-green-walls

NYC. (2019). *Green infrastructure*. New York City Department of Environmental Protection. Retrieved 23 November, 2020, from https://www1.nyc.gov/site/dep/water/types-of-green-infrastructure.page

Provost, C. (2012, 6 September). *Slum surveys giving "invisible" inhabitants a say in urban planning*. The Guardian. Retrieved 23 November, 2020, from https://www.theguardian.com/global-development/2012/sep/07/slum-dwellers-say-in-urban-policy

Ramin, B. (2009). Slums, climate change and human health in sub-Saharan Africa. *Bulletin World Health Organisation*, *87*(12), Article 886.

Rosenzweig, C., Solecki, W. D., Romero-Lankao, P., Mehrotra, S., Dhakal, S. and Ibrahim, S. A. (Eds). (2018). *Climate Change and Cities: Second Assessment Report of the Urban Climate Change Research Network*. Cambridge: Cambridge University Press.

Roy, A. (2005). Urban informality: Toward an epistemology of planning. *Journal of the American Planning Association*, *71*(2), 147–158.

Roy, A. (2009). Why India cannot plan its cities: Informality, insurgence and the idiom of urbanization. *Planning Theory*, *8*(1), 76–87.

Roy, A. (2010). Informality and the politics of planning. In J. Hillier and P. Healy (Eds), *The Ashgate Research Companion to Planning Theory* (pp. 105–126). Abingdon: Routledge.

Saunders, W. S. A. and Becker, J. S. (2015). A discussion of resilience and sustainability: Land use planning recovery from the

Canterbury earthquake sequence, New Zealand. *International Journal of Disaster Risk Reduction*, *14*(1), 73–81.

Savage, M., Dougherty, B., Hamza, M., Butterfield, R. and Bharwani, S. (2009). *Socio-Economic Impacts of Climate Change in Afghanistan*. Stockholm: Stockholm Environment Institute.

Shaw, R., Gupta, M. and Sharma, A. (2005, 27–28 September). Mason's association in Gujarat, India: An innovative approach of linking knowledge and practice in non-engineered construction through government-non-government partnership [Paper presentation]. APEC-EqTAP Seminar on Earthquake and Tsunami Disaster Reduction, Jakarta, Indonesia.

Shaw, R. and Izumi, T. (2014). *Civil Society Organization and Disaster Risk Reduction*. Japan: Springer Verlag.

Siavash, Y. S. (2016). *Achieving urban resilience: Through urban design and planning principles* [PhD thesis]. Oxford Brookes University, UK.

Song, L. K. (2016). Planning with urban informality: A case for inclusion, co-production and reiteration. *International Development Planning Review*, *38*(4), 359–381.

Storbjörk, S. and Uggla, Y. (2015). The practice of settling and enacting strategic guidelines for climate adaptation in spatial planning: Lessons from ten Swedish municipalities. *Regional Environmental Change*, *15*(6), 1133–1143.

Theilacker, J. (2019). *Transfer of Development Rights*. Pennsylvania Land Trust Association.

UN-Habitat. (2015). *International Guidelines on Urban and Territorial Planning*. Nairobi: United Nations Human Settlements Programme.

UN-Habitat. (2016). *Urbanization and Development: Emerging Futures*. Nairobi: United Nations Human Settlements Programme.

UN-Habitat. (2017). *New Urban Agenda*. New York: United Nations.

UN-Habitat. (2020). *Afghanistan – population living in slums (% of urban population)*. Index Mundi. Retrieved 23 November, 2020, from https://www.indexmundi.com/facts/afghanistan/indicator/EN.POP.SLUM.UR.ZS

United Nations. (2019). *Sustainable Development Goals*. Retrieved 23 November, 2020, from https://sustainabledevelopment.un.org/topics/sustainabledevelopmentgoals

United Nations International Strategy for Disaster Reduction. (2005). *Hyogo Framework for Action 2005-2015: Building the Resilience of Nations and Communities to Disasters*. Geneva: United Nations.

United Nations International Strategy for Disaster Reduction. (2015). *Sendai Framework for Disaster Risk Reduction*. Geneva: United Nations.

Velazquez, L. (2019). *New York passes mandatory green roof legislation*. Greenroofs.com. Retrieved 23 November, 2020, from https://www.greenroofs.com/2019/04/18/april-18-2019-new-york-passes-mandatory-green-roof-legislation/

Wang, Y. (2015). Negotiating the farmland dilemmas: "Barefoot planners" in China's urban periphery. *Environment and Planning C: Government and Policy*, *33*(5), 1108–1124.

Weiyuan, C. (2008). China's village doctors take great strides. *Bulletin of the World Health Organization*, *86*(12), 909–988.

World Bank. (2015a). *Building Regulation for Resilience*. Washington DC: World Bank.

World Bank. (2015b). *Urban redevelopment*. Retrieved 23 November, 2020, from https://urban-regeneration.worldbank.org/node/32

World Bank. (2019a). *Master planning*. Retrieved 23 November, 2020, from https://urbanregeneration.worldbank.org/node/5

World Bank. (2019b). *Population living in slums (% of urban population)*. Retrieved 23 November, 2020, from https://data.worldbank.org/indicator/EN.POP.SLUM.UR.ZS

Yamagata, Y. and Sharifi, A. (Eds). (2018). *Resilience-Oriented Urban Planning: Theoretical and Empirical Insights*, Vol. 65. Cham: Springer.

Ziervogel, G. (2019). Building transformative capacity for adaptation planning and implementation that works for the urban poor: Insights from South Africa. *Ambio*, *48*(5), 494–506.

Chapter 5

Resilient urban systems and services: From hard to soft infrastructure

THE global community of practice working to enhance urban resilience has employed urban services and systems as an entry point for improving the ability of towns and cities to withstand climate impacts. This has been done largely through the installation of more robust "hard infrastructure" (i.e. physical systems) to ensure that urban residents are able to access water, energy, transport, communication, public health and emergency services during shocks and stresses. In contrast, investments in "soft infrastructure" (institutional, organisational and individual capacity) have not kept pace (Archer and Dodman 2015; Shakya et al. 2018).

This disproportionate emphasis has led to a vision of change based more on engineering resilience; this considers the resilience of a city as directly dependent on "the capability of all the physical components of the system, including buildings and transportation infrastructures, to absorb the damages due to an external shock and to quickly restore their state before the shock" (UN-Habitat 2017: 5). This conceptualisation overlooks the sociopolitical components and inherent dynamism of urban centres, considering them to be stable systems impacted by predictable disturbances, that can be managed by top-down, technoscientific responses. As such, it stands in contrast to more evolved, social-ecological understandings of urban resilience in which the ability of towns and cities to withstand disturbances is seen as predicated on multiple, interacting factors. The urban resilience framework set out by Tyler and Moench (2012) (explored in Chapter 1) centres on the interaction of resilient systems (physical systems and infrastructure), agents (actors in urban systems) and institutions (rules and behaviours). This chapter argues that here has been an over-reliance on techno-managerial approaches and physical infrastructure, reflecting wider biases in climate adaptation and resilience-building efforts (Nightingale et al. 2020). While recognising the importance of these approaches, we call for a pivot towards their interrelations with soft infrastructure, the agents and institutions.

The relative lack of attention to building individual competencies, organisational capabilities and institutional capacities has been recognised in international policies. In the United Nations Framework Convention on Climate Change, resolutions in support of capacity building in developing countries were adopted in the 2001 Marrakesh Accords and reiterated in the formation of the Paris Committee on Capacity-building (Archer and Dodman 2015; Dagnet et al. 2015; UNFCCC 2019a). The Sustainable Development Goals highlight the need to strengthen capacity for adaptation to climate change, extreme weather, drought, flooding and other disasters along with the capacity for participatory, integrated and sustainable human settlement planning (UNGA 2015).

This chapter begins by exploring the state of practice in building urban systems resilience. It then outlines the challenges of this dominant approach and sets out a pivot towards enhancing soft infrastructure through individual competencies, organisational capabilities and institutional capacities.

5.1 RESILIENT URBAN SYSTEMS: THE STATE OF PLAY

Urban areas are dependent on a range of service delivery systems. This section focuses on four of these to explore how these are employed as entry points for resilience.

5.1.1 Water and sanitation

Climate change will impact both the supply and demand for water in cities. Rising temperatures and erratic rainfall will exacerbate the depletion of groundwater sources through both increased use and the degradation of watersheds (32% of the population in the Asia-Pacific region is dependant on groundwater); rising sea levels and storm surges will potentially increase salt water intrusion into freshwater; urban heat island effects will contribute to greater evaporation from tanks and reservoirs; and extreme events will damage water and sanitation infrastructure (NGWA 2015; Revi et al. 2014; Cromwell et al. 2007). There are three primary ways in which the resilience of the water sector has been conceptualised. Resilience can be achieved by limiting the demand for or increasing the supply of water to reduce stress on the system and create buffers as well as by ensuring that water infrastructure is geared to withstand extreme events (Hatton et al. 2019; James et al. 2018).

There are a number of examples that demonstrate the pathways through which the demand for water has been controlled. In the United States, California has suffered catastrophic drought in the recent past, and the governor declared a "drought state of emergency" in 2014, lifted only in 2017 (USGS 2020). The state has taken numerous measures to reduce the demand for water, including stipulating designs for showerheads and faucets to reduce water flow. In 2015, the state passed an order to reduce by 20% the amount of water that could flow through a showerhead, and three years later, in 2018, this was reduced by another 20% (CEC 2015). Overall, these new changes were expected to save almost 2.4 billion gallons of water in the first year of their implementation and 38 billion gallons over a ten-year period. A similar demand management approach, the Water Efficiency Labelling Scheme, has been applied in Australia (Hatton et al. 2019). By labelling domestic water products with consumption information, authorities have influenced purchasing decisions and driven down demand. As a result, the country is saving 100 gigalitres of water per year.

Along with demand management, augmenting and enhancing supply is another strategy that has been employed to ensure greater resilience in water delivery systems across urban centres. A good example comes from Manila, which is suffering from acute water shortages that are likely to worsen with climate change (WHO 2019). Additionally, the city has historically had an old and inefficient water delivery system with extremely high transmission losses (ADB 2010). Recognising the need to make the city more resilient to water shortages, the city's water utilities undertook a wide range of infrastructural reforms. This included reconfiguring the pipe network into number of smaller hydraulic areas, replacing and laying 3,400 kilometres of new water pipes and employing technology solutions such as correlators and ground microphones for accurate and timely leak detection

(Lindfield and Steinberg 2012). These changes have resulted in a reduction of water losses from 45% in 1996 to 12% in 2013 in the Eastern zone and from 66% to 39% in the Western zone for the same period, leading to improved supply (Verougstraete and Enders 2014).

Reducing the risk of disruption from extreme events is the third pillar of climate-resilient water management. There are numerous examples of this from across the world, but an effective one comes from London, United Kingdom, where the Thames Barrier was built to enhance resilience against storm surges and other kinds of flooding. Storm surge is the "rise in the seawater level during a storm, measured as the height of the water above the normal", and these are likely to worsen with climate change (NOAA 2019; Rahmstorf 2017). Britain has historically been impacted by this phenomenon, that can wreak havoc with the water supply system in an urban centre by destroying supply infrastructure and through storm water intrusion into freshwater reservoirs (Prichard 2013). The Thames Barrier is located downstream from London and consists of 10 gates that are 20 meters high; when closed, it prevents storm surges from coming in from the sea travelling up the Thames, thereby protecting the city and its service delivery systems from flooding.

5.1.2 Energy and power

Climate change is set to impact both the demand and the supply of energy in towns and cities. Supply will be affected in a number of ways; for example, there will be lower hydroelectric power generation due to the depleting volume of water in rivers and more erratic supply from thermal power plants that are dependent on large volumes of water for cooling due to increasing water scarcity (US EPA 2015). Gradual climatic changes such as sea level rise and extreme events carry the potential to destroy the energy supply infrastructure, as a large percentage of the world's oil refineries are located at sea level and on the coast (Revi et al. 2014). On the demand side, extreme temperatures will exacerbate urban heat island effects, leading to enhanced demand for air conditioning. Shortages in regular water supply will be met using pump sets which, in turn, will require additional power supply. Resilience of the energy sector can be ensured by managing demand, augmenting supply and ensuring that energy infrastructure is geared to withstand extreme events.

There are several examples that demonstrate pathways for reducing energy demand. Approaches to help manage demand include spreading awareness of technology solutions, introducing pricing regimes or using other regulatory measures to limit energy use. A good example that combines these different levers is the transition towards use of light-emitting diode (LED) lighting in cities across the world to reduce energy demand. A project run by the Climate Group (an international NGO) collaborated with governments to gauge the impact of a transition to LEDs for street lighting (Climate Group 2012). Working with 12 city governments across three continents over a three-year period, the project provided conclusive evidence that LEDs can deliver a reduction in energy demand by 50% to 70%. Given that almost 20% of the global demand for electricity comes from lighting, this holds the potential to deliver massive gains. These trials have contributed to a number of governments announcing plans for a large-scale rollout of this technology. For example, the Government of India has committed to upgrading 35 million street lights across urban centres to LEDs (Climate Group 2017).

Supply-side approaches have also been employed. Unlike conventional grids that rely on a one-way transmission from the power plant to the consumer, smart grids enable a two-way feedback, to and from the end user (Jerin et al. 2018). Smart grids strengthen supply, including by using of sensors and smart meters to gauge dips and spikes in demand and use. The grids are then able to direct power from areas that have deficits in power use to areas that may require a surplus for any reason, such as battling extreme temperatures. These grids are being rolled out across urban centres in different countries. For instance, 45 states in the Unites States have implemented this approach across towns and cities of different sizes (Lee et al. 2012).

Energy systems are being made more resilient to the impacts of extreme events. Modular energy systems support resilience by disconnecting discrete components from the grid should the need arise (Faure et al. 2017). For example, when New York City experienced power outages and resulting economic losses during Hurricane Sandy in 2021, at one point almost 20% of the city was without power (Compton 2017). In response, the government has initiated construction of 16 microgrids across the city (UGE 2020). These use renewable sources to store electricity in high-tech lithium ion batteries that go online automatically in the event of grid failure.

BOX 5.1 Smart grids for urban resilience

Smart grids are becoming increasingly popular among those wishing to enhance resilience across the world (Stephens et al. 2013). Climate-change-induced extreme events may cause critical components of an electric supply system (e.g. a transformer) to fail. Traditional electricity grids have one-way transmission from the suppliers of electricity to its users. Smart grids on the other hand, permit a two-way feedback between energy utilities and end users through sensors and smart meters placed in their homes and offices. The smart grid permits energy utilities to direct power away from demand-deficient areas to those that may need additional energy for short periods of time, helping maintain critical balance in the system (Lee et al. 2012). This leads to a more efficient transmission of electricity, swifter detection and resolution of outages, reduced operations costs, better integration of diverse energy sources into the grid and improved security (US DoE 2020). Finally, in many urban areas, consumers retain some backup power systems, such as generators, that lie unused most of the time. In times of crisis, smart grids could help bring these redundant energy sources online from areas that are unaffected and help transmit electricity to areas that may need extra supply (US DoE 2020). Smart grids are seeing wide uptake. They are in use across 45 states in the United States, India has established a National Smart Grid Mission to accelerate their deployment, and they are also being deployed across Latin America, Africa and East Asia (Lee et al. 2012; Government of India 2020).

5.1.3 Transport and communication

Climate change will impact transportation and telecommunications infrastructure in innumerable ways. Direct impacts largely include damage to infrastructure from climate-induced extreme events. In countries across the world, vital transport and telecommunications systems are concentrated in towns and cities, and most of these are located in highly exposed locations (i.e. along coasts and rivers). Taking just one example, a 7 meter storm surge would flood half the major highways, railway lines, most airports and all the ports in the Gulf Coast region of the United States (Revi et al. 2014; Ponzi and Iwasaki 2014). Also, storms and high-speed winds will damage above-ground transmission cables, soil erosion from rising sea levels will expose major telecommunication cables and trunk routes, and telecommunication exchanges and base stations may over-heat due to extreme temperatures, to name just a few impacts (Maunsell, 2008).

Transport systems have been made more resilient using engineering and technical solutions that can support robustness (so as to withstand extreme pressure), redundancy (where disruption in part of the network need not result in catastrophic failure) and flexibility (where the configuration of the infrastructure can change in response to circumstances) (Bahadur et al. 2016). Climate resilience of transportation infrastructure can be made more resilient by influencing its design, location and materials used in its construction. A good example of this is found in Ho Chi Minh City, Vietnam. The city is low-lying and experiences flooding that is set to worsen with climate change (Ponzi and Iwasaki 2014). Therefore, the new Mass Rapid Transit system had to be climate-proofed to ensure its effective operation. This included a higher degree of insulation and waterproofing of all electrical and mechanical operating systems to ensure system robustness; the installation of redundant battery-powered electricity supply that comes online should the main supply fail; and flexible tunnel entrances and exits that can be opened and closed swiftly to prevent inundation or facilitate water evacuation.

Communications systems have received similar treatment in urban centers experiencing climate impacts. Communications infrastructure can be made more robust through the use of higher tensile strength steel for the construction of mobile phone towers so that they can withstand higher wind speeds and ice loads as well as by locating server farms and transmission nodes on higher ground to avoid flooding (Venema and Temmer 2017). A wide-ranging review of the reasons for the failure of communications systems in and around the town of Fukushima after the catastrophic earthquake and tsunami in 2011 identified several measures to enhance the resilience of these systems. Recommendations included burying cables underground, building in power redundancies for communications systems through battery and fuel backups, and enhancing geographic spread and distribution of crucial communication nodes and infrastructure (El Khaled and Mcheik 2019).

5.1.4 Health and social services

Climate change will impact health and social services in a few ways. First, the impacts of climate change will potentially destroy or degrade infrastructure to provide health and social services (Bahadur et al. 2016). Second, the sector may be overwhelmed by spikes in demand due to climate-induced extreme events. Third, the sector may be unable to fulfil its mandate due to shifting disease patterns and new health risks that have not been accounted for. Overall, interventions to enhance the resilience of the health sector to climate change impacts have focused either on securing and disseminating information or on improving and enhancing infrastructure for the delivery of healthcare and social services.

Given their pre-existing proclivity for urban heat island effects, towns and cities the world over are experiencing the catastrophic impacts of heatwaves. The risk from heatwaves is being addressed by early warning systems. One such system aims to enhance the resilience of Ahmedabad, India, which has suffered debilitating heatwaves in the past, including one in 2010 that led to almost 1,000 deaths across the city (Jaiswal and Kaur 2017). In response to this, the Municipal Corporation established a heat early warning system to anticipate and respond to heatwave mortality (Shah et al. 2018). The system entails generation of precise temperature forecasts seven days in advance, that are then used to trigger three different grades of early warning (Shah et al. 2018). These are then used to alert the State Disaster Management Authority, hospitals, water providers, energy utilities, school boards, emergency services and community health groups, so that they are able take appropriate action to minimise damage (AMC 2016).

The other point of focus for enhancing the resilience of health and social services is the ability of healthcare infrastructure to withstand the impacts of extreme events and climate-change-induced stresses. The World Health Organisation has underlined the vital importance of climate-resilient health infrastructure through the promulgation of building codes for health facilities that take into account not only current but also projected climate risk (WHO 2015). There are numerous examples of how this has been done. One effective illustration comes from the US city of Homestead, Florida, which was devastated by Hurricane Andrew in 1992 (Guenther and Balbus 2014). Following this, a series of regulations were released requiring wind- and impact-resistant building envelopes, equipment-anchoring systems and emergency generators above surge levels for all new facilities. In addition, regulations now prohibit new hospital construction within a hurricane surge inundation zone and require "all health care projects to adhere to the SLOSH (Sea, Lake, and Overland Surges from Hurricanes) modelling for Category 3 (Saffir-Simpson scale) storms and to set elevations for floors and patient/resident support infrastructure equipment based upon the results" (Guenther and Balbus 2014: 14).

Table 5.1 summarises the systems for resilence that are in place in different sectors.

5.2 CHALLENGES: DISPROPORTIONATE EMPHASIS ON HARD INFRASTRUCTURE

The preceding section illustrates how approaches for enhancing the resilience of urban systems have laid disproportionate emphasis on hard infrastructure while overlooking the need for institutional, organisational and individual capacity. This fact is also evident from an examination of funding flows and financial investments in urban resilience. In one analysis, less than 10% of all funding from salient climate funds for urban resilience went to low-income countries, and most of this was dominated by one large investment in an urban transportation project (Barnard 2015). This is in line with the overall trend of investment in urban resilience where almost 60% of financing by dedicated climate funds for urban projects was to support urban transport infrastructure projects, whereas investments in soft infrastructure and capacity building were marginal.

Looking at this more closely, the Global Environmental Facility (GEF) is one of the world's most important mechanisms for financing resilience interventions. It was established in 1992 and since then has provided close to USD 20 billion in grants and an additional USD 107 billion in co-financing for more than 4,700 projects in 170 countries (GEF 2019). In their salient analysis of the pathways for strengthening the institutional architecture for capacity building on climate change, Dagnet and colleagues (2015) underline the difficulty of isolating exactly how much GEF financing goes towards capacity building, as these initiatives are embedded within others. They note that enabling activities are the purest form of capacity building but highlight that these account for only 2.6% of total GEF funding.

TABLE 5.1 Resilient urban systems: The state of play

Sector	Risks	Resilience actions
Water and sanitation	• Groundwater depletion • Degradation of watersheds • Saltwater intrusion • Surface water depletion through enhanced evaporation	• Reducing demand – e.g. mandating width of showerheads • Increasing supply – e.g. replacing pipes and using technology to detect leaks • Strengthening infrastructure – e.g. reducing storm water intrusion through protective infrastructure
Energy and power	• Lower hydroelectricity generation due to depleting water in rivers • Destruction and degradation of infrastructure • Heightened demand due to rising temperatures	• Reducing demand – e.g. shift to LED street lighting • Increasing supply – e.g. smart grids to direct power to areas with surplus demand from areas with deficits in demand • Strengthening infrastructure – e.g. use of microgrids that maintain supply by delinking from larger grids during shocks and stresses
Transport and communications	• Degradation and destruction of infrastructure • Disruption of services	• Engineering and technical solutions – e.g. climate-proofing transport infrastructure • Enhancing robustness – e.g. use of high tensile steel for mobile phone towers to enhance wind resistance
Health and social services	• Degradation and destruction of infrastructure • Spikes in demand due to climate extremes • Shifting disease patterns	• Early warning systems – e.g. heatwave warning systems to manage spikes in demand • Climate-proofing infrastructure – e.g. establishing building standards for hospitals to enhance robustness

There is a growing recognition of these skewed investment priorities, and one analysis found that annual investment by governments in urban infrastructure to 2025 will be 300 times larger than adaptation funds (Barnard 2015). The implication of this trend is that

> adaptation funds should focus their efforts on improving the extent to which urban resilience upgrading is mainstreamed into local and national policy, planning and regulatory frameworks, and on building related institutional and technical capacities, rather than simply funding defined infrastructure improvement projects that will only ever benefit a tiny proportion of the total number of urban residents likely to be impacted.
>
> (Barnard 2015: 17)

In this way, the analyst seeks to argue that investments in hard infrastructure by governments are unlikely to abate and therefore development finance and official development assistance focused on urban resilience needs to strengthen soft infrastructure across urban centres.

Even when investments in institutional capacity are made, these are to ensure that maximum benefits from hard infrastructure investments are realised through appropriate policy, regulatory and technical frameworks and institutional capacities at different scales necessary for the successful implementation of a particular urban technology or system (Barnard 2015). The Clean Technology Fund and the Pilot Program for Climate Resilience

> provide a majority of their finance for hard infrastructure investments this is usually coupled with some form of technical assistance or capacity building to improve the chances that these are implemented sustainably, although for the [Clean Technology Fund] in particular these aspects constitute a very small part of total project costs.
>
> (Barnard 2015: 17)

A practical example of this is the Asian Sustainable Transport and Urban Development programme, that seeks to build the capacity of Asian cities to develop sustainably. Looking at this more closely, it becomes apparent that this is limited to the development of policies to enhance the integration and use of new modes of transport (such as bus rapid transport systems) into the urban fabric.

This lack of investment and, by extension, lack of attention to the meaningful development of soft infrastructure leads to a number of concomitant problems.

5.2.1 Poor attention to residual risk

A focus on reducing risk and building resilience through an investment in hard infrastructure has led to a growing recognition of the need to understand inherent versus residual risk. Simply put, inherent risk is the degree of risk in a system before any initiatives or controls to reduce risk are installed or activated. Residual risk, on the other hand, is the risk that remains after such controls and initiatives are put in place (Monahan 2008). The United Nations International Strategy for Disaster Risk Reduction applies this concept specifically to disaster-related shocks and stresses to argue that "the presence of residual risk implies a continuing need to develop and support effective capacities for emergency services, preparedness, response and recovery, together with socioeconomic policies such as safety nets and risk transfer mechanisms" (UNGA 2016: 14). The organisation goes on to argue that residual risk needs to be reduced through compensatory disaster risk management, that enhances the social and economic resilience of individuals and societies through institutional strengthening, financial instruments and capacity development. Therefore, while a focus on hard infrastructure leads to a degree of risk reduction, this cannot completely eradicate the possibility that hazards will lead to loss of property and lives.

This is primarily due to the fact that all hard infrastructural solutions for reducing risk operate within a range of certainty, but climate change is likely to alter the magnitude, frequency, return period and spatial distribution of different climate and natural variables (Gallina et al. 2016). This in turn is increasing the likelihood of black swan hazards (Mitchell et al. 2006). Michel-Kerjan et al. (2013: 997) underline the importance of not underestimating "the large uncertainty in the effect of climate change on hazard levels, given that to date it has not been possible to assign probabilities to different climate scenarios". This apart, hard infrastructural solutions are based on a computation of risk through the use of probabilistic risk modelling that seeks to gauge the factors that can enhance exposure of people and assets to hazards, the likelihood of this enhanced exposure and the consequence of this happening (Modarres 2008). However, the variables that determine exposure (e.g. demographic trends, economic activity, policy shifts, security incidents/threats) are as great in scale as they are varied, and models are able unable to accommodate them all (Michel-Kerjan et al. 2013). Therefore, results help planners deal with a part but not the entire spectrum of risk in any location.

There are numerous examples of how hard infrastructure meant to enhance resilience has been overwhelmed by events outside the range of risk for which it was built. A good illustration of this related to Hurricane Katrina in 2005, a black swan event considered so unlikely that protective infrastructure was not geared to deal with it (Nafday 2009). As discussed in Chapter 1, breaching of the flood control infrastructure around New Orleans (e.g. levees) left 80% of the city underwater. Similarly, the outbreak of COVID-19 is an extreme outlier event that overwhelmed healthcare infrastructure across the world. Hard infrastructure can only help reduce risk up to a point, beyond which there is a need for soft infrastructure – individual competencies, organisational capabilities and institutional capacities for dealing with the risk that remains (UNGA 2016).

The disproportionate emphasis on hard systems has arisen for a number of reasons. First, processes of developing physical systems and infrastructure are more linear and less complex than approaches for enhancing soft systems and capacity. This is because capacity building programmes target individuals (who have subjective views), dynamic organisational structures and/or complex and sometimes amorphous institutional structures characterised by entropy (Auster 1974). Second, financing hard infrastructure is easier for a number of reasons; for instance, it is easier to calculate its costs and benefits (Guild and Drilon 2013). Putting a dollar value on returns from investments in complex capacity building initiatives that might span scales of governance and take place over time in multiple geographies and involving different entities is much more complex (Kim and Mollerus 2016). Third, the development and installation of hard infrastructure can at times be politically rewarding, as it is a physical and visceral manifestation of "development" (Burrier 2019). The establishment of a new microgrid network or the installation of new LED street lights are more likely to receive press coverage than the execution of a long-term and gradual organisational and institutional reform agenda. Finally, the approaches discussed in Section 5.1 emanate from planning regimes aligned with Eurocentric epistemological frames that privilege rationalistic and scientific intellectual investigation, resulting in linear engineering solutions to the complex problem of urban risk (Song 2016).

5.2.2 Limited complex systems thinking

Complex systems thinking acknowledges the intrinsic interconnection between the different elements of a system (Ramalingam et al. 2008). More specifically, there might be connections between individual elements of a system or between subsystems, and these connections might be across levels (Ramalingam et al. 2008). This lens for understanding systems underlines that different system components influence each other. The larger and more complex the system, the more non-linear and random this influence (Ramalingam et al. 2008). Urban areas have long been seen as complex systems (Sanders 2008). With the convergence of a number of service delivery systems, economic sectors and cascading socio-economic issues, urban areas have long been studied through this lens (Ruth and Coelho 2007). The World Bank underlines that "cities are complex systems; and, like all systems, a city depends on the smooth functioning of its constituent elements .... Building a resilient city therefore requires a holistic, multi-sectoral, and flexible approach" (2014: 13). This is why, understanding and managing how the different components of a system relate to each other is essential to ensuring resilience, and this requires complex systems thinking, where the interaction of the different elements of an urban system are analysed (Biggs et al. 2015). Reinforcing feedback (e.g. between polluting urban water bodies and the penalties for this) or dampening feedback (e.g. between polluting effluents and their flow into urban water bodies) between different system elements is a key strategy for ensuring resilience (Biggs et al. 2015).

Tyler and Moench (2012) adopt this thinking in their framework for urban climate resilience. They argue that urban resilience emanates from the interaction of resilient systems, agents and institutions. Systems include physical infrastructure and ecosystems that are essential for water, food, energy, transport, shelter and communication for urban residents. Agents are the individuals, households, private sector companies and public sector organisations that exist and operate within an urban centre. Institutions are the formal, informal, implicit and explicit social rules and conventions that structure human behaviour and influence the interaction of systems and agents.

Another conceptualisation of urban resilience arguing that cities are "complex systems" contends that resilience relies on a "complex web of interconnected institutions, infrastructure and information" (Arup 2014). Here, urban resilience is a function of leadership and strategy, health and well-being, economy and society and infrastructure and ecosystems. Yet another approach sees cities "as a system of systems" and proposes ten essential elements for enhancing resilience, spanning governance and finance, planning and preparation and response and recovery (UNDRR 2017). This approach argues that institutional capacity and resilient infrastructure need to work together to ensure resilient cities. Thus there is broad agreement that different elements of an urban centre – spanning both hard and soft systems – need to be understood and included in any approach for enhancing resilience.

The disproportionate focus on the installation of physical systems is antithetical to analysing urban areas as complex systems, goes against the thrust towards engaging multiple urban systems (enshrined in dominant climate resilience frameworks) and is, therefore, an ineffective approach to enhancing the resilience of towns and cities.

5.2.3 Lack of resilience capacity in cities

The disproportionate emphasis on hard infrastructure has led to a clear gap in the very competencies, capabilities and capacities that are crucial for ensuring that urban centres are resilient to the impacts of a changing climate. These can be defined as "the ability to perform functions, solve problems and set and achieve objectives" (Willems and Baumert 2003: 10).

First, those running urban centres lack authority to take appropriate action to help urban systems deal with the impacts of a changing climate. There are many dimensions to this problem and numerous underlying reasons. Across much of the developing world, processes of decentralisation through which urban local bodies have influence over budgets and major policy decisions are fractured, and authority usually rests with governance actors at higher levels in provincial or national capitals. India exemplifies this conundrum as the 74th amendment to the Indian constitution provides greater authority to urban local bodies, providing state governments with a list of mandatory as well as discretionary powers that should be devolved to such bodies (such as municipal corporations) (Mukhopadhyaya and Jayal 2000). However, in reality, most state governments have taken the minimum necessary actions and retained as much power as possible, leading to a severely fractured decentralisation process (Chamaraj 2009). Therefore, urban local bodies continue to have limited agency in determining regime changes, alterations in protocols and substantial shifts in policy/strategy essential for enhancing resilience (Martins and Ferreira 2011). Similar challenges with the authority of urban governments are common in other countries such as China and Bangladesh (Bahadur and Thornton 2015).

Second, those running urban centres lack the awareness and knowledge necessary to take action to enhance resilience of towns and cities to the impacts of climate change (Koop et al. 2017). A commanding survey of 350 municipalities across five continents found substantial gaps in awareness on *how* the climate would impact the functioning of urban centres and *what* could be done to ameliorate the risk of these impacts (Aylett 2015). Of those surveyed, 53% did not understand how local governments could address the issue of climate change, 51% were unaware of the impacts of climate change and did not comprehend how the issue was relevant locally, 40% were aware that the lack of knowledge on climate impacts was a problem and 36% reported significant challenges from lack of awareness among staff about the significance of the issue in general. There were marked regional variations, and those working within African cities were found to have larger gaps in awareness: almost 70% of the staff working in municipalities lacked awareness about the issue in general and lacked understanding of the local impact of climate change. Similar findings emerged from another survey of local government planners in seven Indonesian cities: "half of the respondents felt that the lack of climate awareness and capacity in local government was impeding coordination for effective responses to climate change" (Archer and Dodman 2015: 70). This apart, others have noted that even where city officials might be aware of climate impacts, an adequate understanding of the root causes of risk and vulnerability essential for building resilience was absent (Archer and Dodman 2015).

Third, poor levels of awareness and authority have resulted in lack of capacity across urban centres of the Global South to take appropriate actions to deliver comprehensive improvements in resilience. This can be seen in multiple ways. There are human resource dimensions to this problem, as most towns and cities do not have the required personnel to take on the task of building resilience. Aylett (2015) found that only 4% of cities in his survey had a team dedicated to dealing with urban climate change issues. Only 8% reported having six or more full-time employees with some oversight over these issues, and 23% reported having only a single member of staff charged with overseeing this as part of a broader portfolio of work (e.g. an urban sustainability coordinator).

Apart from human resource gaps, urban centres across the Global South also face challenges related to scarcity of financial resources and financial management skills necessary to effectively allocate the scarce resources available (Shakya et al. 2018). In one survey, lack of funding for implementation of projects and programmes on climate change was reported as a significant challenge by 78% of cities, and 67% reported lack of funding to hire staff to manage such initiatives (Aylett 2015). At a sectoral level, an analysis of resilience in the electricity sector found similar deficiencies. The authors note that despite the substantial current and expected impacts of climate change on this sector, electricity utilities do not have a resilience officer; they also state that a "key limiting factor for utility resilience is the ability to fund future resilience for existing and planned assets" (Hall et al. 2019: 35).

5.3 PIVOTING TOWARDS SOFT INFRASTRUCTURE FOR URBAN RESILIENCE

Having analysed the state of play on urban systems resilience and the challenges with the existing focus on hard systems, we argue for a pivot towards emphasising soft infrastructure for building resilience. The section explores the nature of individual competencies, organisational capabilities and institutional capacities that are needed to enhance the authority, awareness and ability to take appropriate action for enhancing urban resilience (Willems and Baumert 2003; Shakya et al. 2018; Dagnet et al. 2015). Vandeveer and Dabelko (2001: 20) argue that "all three dimensions (human resources, organisations, and institutions) are interrelated, so initiatives court failure when they largely ignore one or two of the dimensions".

5.3.1 Individual competencies

There is a need to enhance the authority of key actors to conceptualise and execute pathways for enhancing resilience (Willems and Baumert 2003). This can be done in numerous ways, including by embedding urban resilience roles and functions into individual job descriptions in key sectors. A good example of this comes from New Jersey, a US state that is almost entirely urban. Here, as part of a statewide push to develop a systematic response to climate change, the responsibility of developing and executing a strategy for climate change resilience was officially added to the responsibilities of the Assistant Commissioner of the Environmental Protection Department through the promulgation of an executive order (State of New Jersey 2019). This was important for a number of reasons, including the fact that it gave the Assistant Commissioner authority to convene an Interagency Council on Climate Resilience. Comprised of representatives of 16 state agencies, the council aims to develop action plans for holistic, long-term mitigation, adaptation and resilience measures in line with the Statewide Climate Change Strategy; this in turn, helps bring a vision of complex systems thinking to life. In this way, through executive fiat, New Jersey has ensured that one individual is authorised to lead the resilience agenda across the state, and the cross-government group of 16 key individuals are mandated to consider climate change in the context of the individual service delivery systems that they oversee.

Apart from authority, the preceding section also underlined gaps in awareness of key actors in urban centres on various aspects of climate-resilient development (Aylett 2015). Very broadly, individuals must be able to understand *how* the climate is changing and the implications of these changes for the sectors and systems they are overseeing as well as *what* they can do to ameliorate the risk of climate impacts. This translates into the need for basic understanding of climate information and the implications of expected climate changes on key urban sectors. While it is unrealistic to assume that those running urban water, energy, transport and health departments can ever become experts at developing or deciphering climate models, a basic capacity to analyse historical climate data or the projections emanating from models needs to be built across urban centres.

A wide-ranging review of resilience within the electricity and power generation sector mentions this as a key gap, underlining the absence of ability to assess damage from a range of hazards on different types of electric utilities (Hall et al. 2019). A similar review for the railway sector highlights the importance of enhancing awareness through "workshops, guidance, case studies, role specific training and including objectives in an individual's appraisals" (Reeves et al. 2019: 48). Due to the largely compartmentalised nature of urban governments, it is vital to understand the differential impact that climate change may have on individual sectors to ensure appropriate sectoral responses to these impacts (Bahadur and Tanner 2014a).

This apart, as discussed in Section 5.2.3, a majority of those working within urban local bodies did not understand the actions that actors at their level could take to tackle climate impacts (Aylett 2015). Therefore, an awareness of actions and pathways for enhancing resilience also needs to be built. As part of this, it is crucial for those running urban systems and services to understand the differential impacts of climate change on diverse social groups. This is because there is now decades of evidence that demonstrates how women and other marginalised groups (e.g. those from particular castes, the handicapped or migrants) are more vulnerable than others to climate risk (UNFCCC 2019b; Onwutuebe 2019).

Along with raising the authority and awareness of key actors that manage urban service delivery systems, it is vital that their capacity to execute actions to enhance resilience is improved. This includes the ability to assess climate risk at the city or sector scale, a process that begins by reviewing the existing knowledge on climate risk (Dixit 2012; UNFCCC 2020; Bahadur et al. 2015). Depending on available resources, this can also be done by undertaking primary research by generating scenarios, developing models and speaking to experts. The ability to design and plan actions to reduce exposure, vulnerability and hazards (based on the results of the climate risk assessment) comes next (Tonmoy et al. 2019). It is then imperative that designated individuals can cost climate actions, prioritise actions needed and secure the resources required (Dixit 2012). This can be done by reviewing the cost of similar resilience building actions undertaken by others, using commonly available cost estimating tools/databases or through the use of parametric software to determine the material needs of design interventions and then pricing these (NOAA 2013). Key to effective implementation is the continual monitoring of climate risk as well as the effectiveness of resilience-enhancing actions implemented (Willems and Baumert 2003; Shakya et al. 2018). This can be done by measuring economic loss and mortality or morbidity due to climate impacts over time, tracking resilience-enhancing inputs (e.g. delivery of flood protection infrastructure) or by gauging how urban systems perform in the face of shocks and stresses (e.g. length of power outages during hurricanes) (ODI 2015).

There are burgeoning examples of how these competencies can be developed. These approaches broadly fall into those that are didactic and those that are relational. A good example of competency development in former category is the training on building urban climate resilience organised by The Energy Resources Institute, India, for designated individuals from urban local bodies in the state of Odisha (TERI 2015). The training aimed to bridge the science–policy divide by informing city- and state-level decision-makers about the application of climate change science in urban development planning and decision-making. The agenda included a primer on climate impacts in cities, an introduction to climate modelling and risk assessments, approaches for enhancing resilience through mainstreaming into existing sectors, and sources of financing for urban resilience. To incentivise participation, the training was packaged as a certificate course.

The shared learning dialogue approach is an example of a more relational model of developing competencies that entails a two-way exchange of views and information between those leading competency development initiatives and those that are its targets (see Box 5.2).

5.3.2 Organisational capabilities

To ensure that individual actors have the authority to take the necessary steps for enhancing urban resilience, it is crucial that organisations with a role in managing urban service delivery systems have the right capabilities (Shakya et al. 2018). An organisational culture of resilience needs to be created that empowers actors such as the Assistant Commissioner of the Environmental Protection Department (see Section 5.3.1) to take pivotal steps towards enhancing resilience. This can be done in a number of ways, such as embedding resilience within the charter, mandate or high-level operational guidelines of key organisations. For instance, India's main urban development initiative, the Smart Cities Mission, includes risk reduction as one of six key principles within its guidelines (MoUD 2015). The precise modalities through which resilience is to be enhanced are not fleshed out, but this at least provides an entry point for actors to take action that is locally tailored and contextually relevant.

This apart, the role of leadership and champions cannot be discounted in approaches for enhancing organisational

BOX 5.2 Capacity building for urban resilience through shared learning dialogues

Climate risk in cities is dynamic, stems from multiple and interacting root causes, is without precedent and requires a diversity of stakeholders to formulate a comprehensive response (Camillus 2008). Understanding how diverse service delivery systems, economic sectors and cascading socio-economic issues interact is key to enhancing urban resilience (Meerow and Newell 2015). Relational approaches that bring together the expertise, knowledge and viewpoints of diverse stakeholders in an iterative and structured manner are key for the co-creation of knowledge for enhancing urban resilience.

The shared learning dialogue is one such tool that has been employed to build capacity for urban resilience. This is based on the contention that "by fostering iterative deliberation, sharing of sector- or group- specific knowledge and knowledge from both local practitioners and external experts, the quality and effectiveness of decision-making will be improved" (ISET 2010: 11). Shared learning dialogues are evolved participatory meetings that are cyclical; engage stakeholders from across sectors (to extend understanding of complex interlinkages between sectors), disciplines and scales; aim for the co-production of resilience pathways; emphasise learning by doing; and promote an appreciation of complexity and uncertainty (Reed et al. 2013).

In Gorakhpur, India, shared learning dialogues were employed to bring together low-income communities and policymakers to jointly to understand the underlying drivers of flood risk that were found to be political and not purely biophysical (as was previously assumed) (Bahadur and Tanner 2014b). In Indonesia, this approach strengthened the capacity of non-governmental actors and academics to engage with government on resilience policymaking. This helped bring new perspectives to the table that permitted an examination of the complex factors that underly urban climate risk. In the Vietnamese cities of Quy Nhon and Da Nang, the participation of a variety of stakeholders from disparate sectors led to a gradual understanding of how "uncoordinated urban development had exacerbated flood risk", making way for effective policy measures to enhance resilience (Reed et al. 2013: 404).

capabilities for urban climate change resilience (Leck and Roberts 2015). A recent review of resilience in the water sector underlines the importance of "champions who run scenarios and exercises to test responses to potential failures and learn lessons to reduce occurrences of crises" (Hatton et al. 2019: 26). Also called policy entrepreneurs, champions can help "to gain attention and support for particular issues and possible solutions, for the creation of coalitions in which ideas and policies can be developed and implemented, and in influencing the time and place where decisions are made" (Beunen and Patterson 2019: 19). In the case of New Jersey, the governor personally championed the climate agenda, which in turn has led to an enabling environment for actors such as the Assistant Commissioner to take necessary action (NYT 2018).

Along with authority, it is imperative that organisations have an appropriate amount of information, knowledge and awareness to take relevant action to support urban resilience. There is consensus on the need to move away from models of organisational learning that aim to predict and optimise the future; instead there is a growing emphasis on enabling learning that supports adaptive management (Shakya et al. 2018). This is a style of management that entails an "iterative process for continually improving management by learning from how current management affects the system" (Bunnefeld et al. 2015: 1). This is a particularly important capability for organisations that oversee the management of service delivery systems, because they need to constantly watch how climate change is impacting these systems and whether measures taken to ameliorate this impact are working. This model of management is implemented by the Santa Clara Water Valley District (SCWVD), California – one of the oldest water suppliers in this US state (Conrad et al. 2019). In order to manage drought risk, the SCWVD is mandated to maintain a certain amount of buffer capacity in underground aquifers, and to ensure this, the agency issues mandatory conservation targets to downstream users. In times of crisis, the SCWVD adopts an adaptive management approach, tracking water levels closely and issuing conservation targets in rapid cycles that lead to a frequent increase or decrease in water demand to ensure a delicate balance between user needs and the overall resilience of the system. Similar adaptive planning techniques have been adopted by a number of Australian water utilities, that have identified trigger points for swiftly switching from one option to another to maintain water supply (Hatton et al. 2019: 30). Interestingly, during the COVID-19 pandemic, the manner in which public spaces and transport systems across the world have opened and closed in concert with the rise and fall in infection rates has been seen by some as reactive adaptive management. Proactively enhancing the ability of those managing urban systems to embrace this approach is likely to result in a higher degree of resilience to multiple, interacting shocks and stresses.

In addition, organisational capabilities need to be built to ensure that individuals are able to take adequate and appropriate action to enhance urban resilience. As discussed in Section 5.2, complex systems thinking is crucial for ensuring urban resilience, because feedbacks between different system elements need to be understood to determine the best possible method for managing for resilience. Operationally, this means that organisations charged with overseeing particular urban service delivery systems need to have the capability to engage, liaise and collaborate with one another. A recent review of resilience in the water sector makes this point, noting that there is "a need to consider resilience at a system level. For example, where a large urban centre has multiple organisations involved in water supply, the collective responses of the organisations is as important as the individual components" (Hatton et al. 2019: 14).

The need for cross-sectoral platforms sounds intuitive but is not commonplace. Bureaucratic contexts across the world, especially in developing countries, are notoriously compartmentalised, and organisational processes need to be instituted to facilitate joined-up decision-making (Littler 1978; Weber 2009; Welp et al. 2007). Two good examples of how this can be achieved in the transport sector are found in the United Kingdom where the Department for Transport recently established a cross-modal resilience team and Transport Scotland set up the Multi-Agency Response Control Centre, which brings together key partners from Transport Scotland, the Met Office, ScotRail, operating companies and Police Scotland (Reeves et al. 2019). In another example, in Da Nang city, Vietnam, the Climate Change Coordination Office plays precisely this role and facilitates collaboration between agencies and sectors to enhance resilience.

The ability to access and manage climate finance is also a crucial organisational capability, and "identifying funding sources for adaptation planning and subsequent implementation is a challenge for many city governments" (Hughes 2015: 23). Key to building this capability is strengthening financial and fiduciary management systems that then permit international donors or the private sector to make investments or award grants for enhancing resilience. For instance, the Ministry of Water and Environment, Uganda – the lead organisation charged with managing water supply across rural and urban areas of the country – went through an intensive process of certification to become an accredited entity to secure international climate finance for enhancing resilience from the Green Climate Fund (GCF 2020).

A number of different approaches are being employed to enhance organisational capabilities for supporting urban resilience across the world. Technical assistance initiatives, where external experts work closely with organisations overseeing urban systems to enhance organisational processes and transfer knowledge, have proven effective in different contexts (Shakya et al. 2018). For instance, in Nepal, an initiative funded by the UK government has facilitated a close and long-term engagement between international experts and the country's Ministry of Urban Development. This led to the formulation and institution of a risk and resilience review protocol that ensures that all designs for new urban infrastructure are screened to ensure that they take climate risks into account.

Similarly, organisational certification programmes have demonstrated value beyond the realm of finance. For instance, the World Bank-monitored Excellence in Design for Greater Efficiencies (EDGE) certification is provided to buildings that integrate features to reduce water and energy use by 20%, and this is incentivising more climate-friendly construction across the world (IFC 2020). For instance, Bancolombia, Colombia's largest bank gives discounted loans for buildings with plans that meet the EDGE standard (Menes 2019b). In India, city governments allow EDGE-certified buildings a larger Floor Service Index (the amount of construction permitted per unit of land) (Menes 2019a). In Romania, those purchasing EDGE-certified homes benefit from more attractive mortgages (Menes 2017). Almost 100,000 homes have been registered for this certification across the world (IFC 2020). In this way, certification programmes can create a standard which can then be used to devise an array of incentives for low-carbon climate-resilient development.

Before moving onto the next section, it is important to recognise that processes of enhancing institutional capabilities are not linear. There is a substantial and growing literature on the role of champions, policy entrepreneurship, shadow systems and coalitions for change that are pivotal in processes of improving or shifting organisational processes and protocols

(Leck and Roberts 2015; Willems and Baumert 2003; Roberts and King 1991; Armijos et al. 2017; Pelling et al. 2008). Tanner et al. (2019) provide examples from South Asia of how technical assistance for enhancing organisational capabilities achieved success because those leading these initiatives used trusted intermediaries, employed informal influencing approaches and tagged resilience onto existing policy priorities to elicit interest and engagement from government organisations.

5.3.3 Institutional capacities

An effective institutional environment is essential for ensuring that organisations are authorised to take decisions to enhance urban resilience (Lankao and Gnatz 2015). As discussed earlier in this chapter, a major impediment for this is fractured processes of decentralisation where government organisations overseeing urban service delivery systems lack the agency to take appropriate actions to reduce risk and enhance resilience. Therefore, institutional shifts are needed to remedy this. For instance, Nepal's National Adaptation Programme of Action, endorsed in 2010, stipulated that 80% of the available budget for adaptation will be earmarked for local implementation of identified adaptation actions in municipalities and villages. This led to the creation of Local Adaptation Plans of Action through which local governments are authorised to design adaptation actions and allocate resources (Chaudhury et al. 2014). Apart from decentralising financing and resilience planning functions, it is vital that an institutional architecture exists for mainstreaming or embedding resilience within particular projects and programmes. For instance, the government of Punjab province, Pakistan, passed an order that every publicly funded project with a budget of over 1 billion Pakistani rupees be put through a risk screening and resilience mainstreaming protocol (Acclimatise 2019).

Similarly, an enabling institutional environment needs to be built to ensure that organisations and individuals have the requisite information and knowledge on which to base actions for enhancing urban resilience. This could be done in several different ways, including mandating particular organisations to lead the generation and dissemination of climate information and data. For instance, in the United Kingdom, this mandate lies with the country's Met Office, which leads the UK Climate Projections programme that produces the climate information on which adaptation and resilience plans are based (Met Office 2020). Taking another example, New York City passed a law to give this mandate to the New York Panel on Climate Change (a semi-autonomous body of research scientists), which is charged with providing "an authoritative source of actionable information on future climate change and its potential impacts to support City decision-making" (NYC Mayor's Office of Resiliency 2020).

Institutional architecture can also be created to build the capacity of government personnel as well as non-government actors working in key sectors. For example, South Africa's National Skill Development Strategy lays major emphasis on the development of "green skills" (i.e. jobs that help to protect ecosystems and biodiversity and reduce energy, materials and water consumption) (DHET 2013). The country also has a national Environmental Sector Skills Plan, which lays out a strategic and systematic vision for enhancing the number and quality of personnel able to ensure climate-compatible development across sectors (DEA 2010). Additionally, it is important to acknowledge how institutional factors can enable or impede adaptive management. Frohlich et al. (2018) empirically demonstrate the ways in which rigid legal frameworks that govern environmental management processes in parts of the United States and Canada pose tangible obstacles for adaptive management and argue for more flexibility in these institutional norms.

Action for enhancing resilience in towns and cities needs to stem from a comprehensive and consolidated policy or strategy that allocates responsibilities and provides direction. Unfortunately, most urban centres across the world (especially in developing countries) lack such instruments. This in turn means that they are unable to determine their individual courses of action for urban resilience independent of other levels of government and "implies that local decision-makers are dependent on regional, national, and international regulatory umbrellas that provide incentives and resources for cities to undertake large-scale climate action" (Fuhr et al. 2018: 3). Certain initiatives have attempted to plug this gap; for instance, the 100 Resilient Cities programme supported by the Rockefeller Foundation has built the capacity of city-level stakeholders across the world to consolidate "city resilience strategies" that list priority resilience building actions and road maps for executing these. Apart from enhancing the policy architecture, the "rules of the game" need to ensure that adequate incentives are in place for enhancing action on urban resilience. A good example of how this can be done comes from India, where there is a federal structure and therefore the national government shares tax revenues with provincial governments (Ramalingom and Kurup 1991). A complex formula with a number of performance-oriented parameters helps determine the share that each provincial government receives. Recently "risk mitigation" (including capacity building) has been included as one such parameter, and progress on this will contribute to determining the scale of financial transfers from the federal level to each provincial government (Vatsa 2019).

Numerous approaches have been taken to affect these institutional shifts to enhance capacity for urban resilience. Transnational municipal networks that link cities to one another have delivered some of these shifts, and these networks "have become notable actors in advancing innovative planning strategies such as encouraging cross-sectoral learning, translating science, harnessing new funding mechanisms, and emphasising projects that promote co-benefits" (Bellison and Chu 2019: 76). For instance, the 100 Resilient Cities programme links cities in the Global South with those in the Global North with the intention of inducing an exchange of ideas and knowledge. C40 is another such global network, servicing 96 cities across the world. It offers the opportunity for cities to learn from each other across a number of topics, prominently adaptation and resilience. The network claims that 30% of all actions that member cities have taken to battle climate change resulted from a collaboration with another city in the network, and 70% of all the cities that are part of the network "have implemented new, better or faster climate actions as a result of participating in C40 networks" (C40 2020b). For example, New York City based its strategic approach to dealing with cloudbursts and extreme rainfall on models, approaches and techniques employed by Copenhagen, Denmark, which has a history of responding to this hazard (C40 2020a). Such examples of specific enhancements in institutional capacity as a result of interaction between cities in the network abound. Other networks through which urban centres can receive support and inspiration for enhancing institutional capacity for urban resilience are the ICLEI – Local Governments for Sustainability; United Cities and Local Governments; and the Global Covenant of Mayors for Climate & Energy.

Achieving institutional shifts and enhancements in authority, awareness and ability to take action entails an engagement with political economy. It would be foolhardy to imagine that achieving these enhancements is simply a matter of providing greater access to knowledge and resources. There is now decades of evidence around how complex issues such as incentives, power, values and biases play an important role (Willems and Baumert 2003). A growing number of examples demonstrate this. For instance, Bahadur and Tanner (2014a), in an empirical analysis of an urban context where such institutional shifts for resilience were being effected, found that success depended on coupling capacity enhancement for urban resilience with existing policy priorities, understanding incentives and disincentives of key policy actors and structuring decision-making spaces where key stakeholders felt genuinely included.

Table 5.2 summarises ways that individual competencies, organisational capabilities and institutional capacities are enhanced for urban climate resilience.

5.4 CONCLUSION

This chapter began with an exploration of how major international agreements on climate change and sustainable development emphasise the vital importance of capacity development for resilience. It then examined how, across different contexts, there has been a disproportionate emphasis on strengthening hard systems, with soft systems being relatively overlooked. The analysis outlined how this lack of investment in building capabilities, competencies and capacities has led to a degree of residual risk. Finally, examples were provided from across the world that demonstrate what it takes to develop the soft infrastructure that supports resilient urban systems and services.

Sustained and structured programmes of capacity building have been stressed repeatedly in this field (Dagnet and Northrop 2015; Dagnet et al. 2015; Shakya et al. 2018). Empirical evidence points to the ineffectiveness of ephemeral, ad hoc and piecemeal initiatives and emphasise the vital of importance long-term efforts that adopt a systemic perspective for strengthening soft infrastructure. The importance of enhancing competencies, capabilities and capacities of individuals, organisations and institutions beyond those that lie at the local or city level comes up time and again (Barnard 2015; Archer and Dodman 2015). Efforts at strengthening soft infrastructure at the city level will not be effective or sustainable unless this is buttressed by corresponding shifts at the provincial and national levels. Much of this chapter has focused on enhancing the capacity of government actors, organisations and institutions, as these are the primary catalysts of resilience across urban centres of the Global

TABLE 5.2 Enhancing competencies, capabilities and capacities for urban climate resilience

	Authority	Awareness	Action	Modalities
Individual competencies	• Embedding urban resilience in terms of reference • Empowering ability to convene cross-sectoral stakeholders	• Enhancing understanding of climate change and pathways of resilience • Raising awareness of differential impacts of climate change	• Increasing ability to undertake situation analysis, cost and plan climate actions and monitor changes in resilience	• Didactic training • Relational learning
Organisational capabilities	• Embedding resilience within the charter, mandate or guidelines • Finding champions	• Enabling a shift from learning that aims to predict and optimise to iterative learning that supports adaptive management	• Supporting systems thinking within organisations • Strengthening financial systems to improve climate finance management potential	• Technical assistance • Certification programmes • Understanding approaches for influencing
Institutional capacities	• Ensuring devolved decision-making • Securing a mandate for mainstreaming	• Designating institutional responsibility for producing and disseminating climate information • Instituting systems for structural capacity building on urban resilience • Ensuring a flexible regulatory framework to support adaptive management	• Enabling the development of resilience strategies • Structuring institutional incentives for action on urban resilience	• Engaging in transnational city networks • Understanding institutional politics and political economy issues

South (Archer and Dodman 2015). Yet there is wide recognition of the role that the private sector, civil society organisations and urban citizens need to play in creating a supportive ecosystem for urban resilience (Hughes 2015; UN-Habitat 2017). Initiatives that aim to enhance authority, awareness and action must consider these vital stakeholders going forward.

Finally, it is crucial to recognise that this chapter in no way argues for ceasing the investment of time, attention and finance into the hard infrastructure needed to make urban systems and services more resilient to climate impacts. Instead, it offers a clarion call to balance this with investments in soft infrastructure. Strong systems that are able to withstand disturbances, combined with individuals, organisations and institutions that have the authority, awareness and ability to act, will ensure that marginalised urban populations are able to secure the basic services necessary to function and flourish despite the exigencies of a changing climate.

REFERENCES

Acclimatise. (2019). *Punjab's pact on climate resilience*. Acclimatise News. Retrieved 23 November, 2020, from https://www.acclimatise.uk.com/2019/03/25/punjabs-pact-on-climate-resilience/

Ahmedabad Municipal Corporation. (2016). *Ahmedabad Heat Action Plan*. Ahmedabad: Ahmedabad Municipal Corporation.

Archer, D. and Dodman, D. (2015). Making capacity building critical: Power and justice in building urban climate resilience in Indonesia and Thailand. *Urban Climate*, *14*(1), 68–78.

Armijos, M. T., Phillips, J., Wilkinson, E., Barclay, J., Hicks, A., Palacios, P., Mothes, P. and Stone, J. (2017). Adapting to changes in volcanic behaviour: Formal and informal interactions for enhanced risk management at Tungurahua Volcano, Ecuador. *Global Environmental Change*, *45*, 217–226.

Arup. (2014). *The City Resilience Framework*. London: Arup.

Asian Development Bank. (2010). *Country water action: Winning the war against leaks and losses*. Retrieved 23 November, 2020, from https://www.adb.org/results/country-water-action-winning-war-against-leaks-and-losses

Auster, R. D. (1974). The Gpitpc and institutional entropy. *Public Choice*, *19*, 77–83.

Aylett, A. (2015). Institutionalizing the urban governance of climate change adaptation: Results of an international survey. *Urban Climate*, *14*(1), 4–16.

Bahadur, A., Jobbins, G., Grist, N. and Allison, C. (2015). *Integrating Disaster Risk Reduction, Environment and Climate Change Adaptation and Mitigation into Australian Aid Projects, Programmes and Investments*. London: Overseas Development Institute.

Bahadur, A. and Tanner, T. (2014a). Policy climates and climate policies: Analysing the politics of building urban climate change resilience. *Urban Climate*, *7*, 20–32.

Bahadur, A. and Tanner, T. (2014b). Transformational resilience thinking: Putting people, power and politics at the heart of urban climate resilience. *Environment and Urbanization*, *26*(1), 200–214.

Bahadur, A., Tanner, T. and Pichon, F. (2016). *Enhancing Urban Climate Change Resilience: Seven Entry Points for Action*. Manila: Asian Development Bank.

Bahadur, A. and Thornton, H. (2015). Analysing urban resilience: A reality check for a fledgling canon. *International Journal of Urban Sustainable Development*, *7*(2), 196–212.

Barnard, S. (2015). *Climate Finance for Cities: How Can International Climate Funds Best Support Low-Carbon and Climate Resilient Urban Development*. London: Overseas Development Institute.

Bellinson, R. and Chu, E. (2019). Learning pathways and the governance of innovations in urban climate change resilience and adaptation. *Journal of Environmental Policy & Planning*, *21*(1), 76–89.

Beunen, R. and Patterson, J. J. (2019). Analysing institutional change in environmental governance: Exploring the concept of "institutional work". *Journal of Environmental Planning and Management*, *62*(1), 12–29.

Biggs, R., Schlüter, M. and Schoon, M. L. (Eds). (2015). *Principles for Building Resilience: Sustaining Ecosystem Services in Social-Ecological Systems*. Cambridge, UK: Cambridge University Press.

Bunnefeld, N., Redpath, S. and Irvine, J. (2015). *A Review of Approaches to Adaptive Management*. Scottish Natural Heritage Commissioned Report No. 795. Inverness: Scottish National Heritage.

Burrier, G. (2019). Politics or technical criteria? The determinants of infrastructure investments in Brazil. *The Journal of Development Studies*, *55*(7), 1436–1454.

C40. (2020a). *Cities100: New York City and Copenhagen – cities collaborating on climate resilience*. Retrieved 23 November, 2020, from https://www.c40.org/case_studies/cities100-new-york-city-and-copenhagen-cities-collaborating-on-climate-resilience

C40. (2020b). *Cities will shape our future*. Retrieved 23 November, 2020, from https://www.c40.org/networks

California Energy Commission. (2015). *Energy Commission approves new standards*. Retrieved 23 November, 2020, from https://energyarchive.ca.gov/releases/2015_releases/2015-08-12_approval_new_water_standards_nr.html

Camillus, J. C. (2008). Strategy as a wicked problem. *Harvard Business Review*, *86*(5), 98.

Chamaraj, K. (2009). *Parastatals and task forces: The new decision-makers*. India Together. Retrieved 23 November, 2020, from www.indiatogether.org/2009/feb/gov-parastate.htm

Chaudhury, A.S., Sova, C.A., Rasheed, T., Thornton, T. F., Baral, P. and Zeb, A. (2014). *Deconstructing Local Adaptation Plans for Action (LAPAs): Analysis of Nepal and Pakistan LAPA Initiatives*. Working Paper No. 67. CGIAR Research Program on Climate Change, Agriculture and Food Security. Copenhagen: CCAFS.

Climate Group. (2012). *Lighting the Clean Revolution: The Rise of LEDs and What it Means for Cities*. United Kingdom: The Climate Group.

Climate Group. (2017). *Driving Energy Efficiency and City Renovation*. United Kingdom: The Climate Group.

Compton, A. (2017). *Hurricane Sandy New York City power outage map: Thousands without electricity in metro area*. Huffpost. Retrieved 23 November, 2020, from https://www.huffingtonpost.in/entry/hurricane-sandy-new-york-city-power-outage-map_n_2050380?ri18n=true

Conrad, E., Moran, T., Crankshaw, I., Blomquist, W., Martinez, J. and Szeptycki, L. (2019). *Putting Adaptive Management into Practice: Incorporating Quantitative Metrics into Sustainable Groundwater Management*. Stanford Digital Repository. https://purl.stanford.edu/hx239rw5017

Cromwell, J. E., Smith, J. B. and Raucher, R. S. (2007). *Implications of Climate Change for Urban Water Utilities*. Washington DC: Association of Metropolitan Water Agencies.

Dagnet, Y. and Northrop, E. (2015). *3 Reasons Why Capacity Building Is Critical for Implementing the Paris Agreement*. Washington DC: World Resources Institute.

Dagnet, Y., Northrop, E. and Tirpak, D. (2015). *How to Strengthen Institutional Architecture for Capacity Building to Support the Post-2020 Climate Regime*. Washington DC: World Resources Institute.

Department for Higher Education and Training. (2013). *National Skills Development Strategy III*. Retrieved 23 November, 2020, from www.dhet.gov.za/Booklets/NSDS%20III%20Progress%20Report%20-%207%20October%202013%20-%20V11.pdf

Department of Environmental Affairs. (2010). *Environmental Sector Skills Plan for South Africa*. Pretoria: Department of Environmental Affairs South Africa.

Dixit, A. (2012). *Ready or Not: Assessing Institutional Aspects of National Capacity for Climate Change Adaptation*. Washington DC: World Resources Institute.

El Khaled, Z. and Mcheick, H. (2019). Case studies of communications systems during harsh environments: A review of approaches, weaknesses, and limitations to improve quality of service. *International Journal of Distributed Sensor Networks*, *15*(2), Article 1550147719829960.

Faure, M., Salmon, M., El Fadili, S., Payen, L. and Kerlero, G. (2017). *Urban Micro-Grids*. Paris: ENEA Consulting.

Frohlich, M. F., Jacobson, C., Fidelman, P. and Smith, T. F. (2018). The relationship between adaptive management of social-ecological systems and law. *Ecology and Society*, *23*(2), Article 23. https://doi.org/10.5751/ES-10060-230223

Fuhr, H., Hickmann, T. and Kern, K. (2018). The role of cities in multi-level climate governance: Local climate policies and the 1.5°C target. *Current Opinion in Environmental Sustainability*, *30*, 1–6.

Gallina, V., Torresan, S., Critto, A., Sperotto, A., Glade, T. and Marcomini, A. (2016). A review of multi-risk methodologies for natural hazards: Consequences and challenges for a climate change impact assessment. *Journal of Environmental Management*, *168*, 123–132.

Global Environmental Facility. (2019). *About us*. Retrieved 23 November, 2020, from https://www.thegef.org/about-us

Government of India. (2020). *National smart grid mission*. Retrieved 23 November, 2020, from https://www.nsgm.gov.in/

Green Climate Fund. (2020). *Ministry of Water and Environment, Uganda*. Retrieved 23 November, 2020, from https://www.greenclimate.fund/ae/mwe-uga?inheritRedirect=true&redirect=/how-we-work/tools/entity-directory

Guenther, R. and Balbus, J. (2014). *Primary Protection: Enhancing Health Care Resilience for a Changing Climate*. Washington, DC: US Department of Health and Human Services.

Guild, R. and Drilon, M. (2013). *Economics of Climate Proofing at The Project Level*. Manila: Asian Development Bank.

Hall, P., Vanderbeck, R. and Triano, M. (2019). *Electric Utilities: An Industry Guide to Enhancing Resilience*. London: Wood Group PLC and Resilience Shift.

Hatton, T., Kay, E., Naderpajouh, N. and Aldrich, D. (2019). *Potable Water: An Industry Guide to Enhancing Resilience*. London: Resilience Shift.

Hughes, S. (2015). A meta-analysis of urban climate change adaptation planning in the US. *Urban Climate*, *14*(1), 17–29.

International Finance Corporation. (2020). *Capitalize on the opportunity to excel with green buildings*. Retrieved 23 November, 2020, from https://www.edgebuildings.com/

ISET. (2010). *The Shared Learning Dialogue: Building Stakeholder Capacity and Engagement for Resilience Action.* Climate Resilience in Concept and Practice Working Paper Series. Boulder, CO: ISET.

Jaiswal, A. and Kaur, N. (2017). *5 reasons why the Ahmedabad Heat Action Plan saves lives.* Natural Resources Defence Council. Retrieved 23 November, 2020, from https://www.nrdc.org/experts/anjali-jaiswal/5-reasons-why-ahmedabad-heat-action-plan-saves-lives

James, A. J., Bahadur, A. V. and Verma, S. (2018). *Climate-resilient water management: an operational framework from South Asia.* Action on Climate Today Learning Paper. New Delhi: ACT.

Jerin, A. R. A., Prabaharan, N., Kumar, N. M., Palanisamy, K., Umashankar, S. and Siano, P. (2018). Smart grid and power quality issues. In H. Fathima, N. Prabaharan, K. Palanisamy, A. Kalam, S. Mekhilef and J. Justo (Eds), *Hybrid-Renewable Energy Systems in Microgrids* (pp. 195–202) Woodhead Publishing.

Kim, N. and Mollerus, R. (2016). *Cost-Benefit Analysis for Identifying Institutional Capacity Building Priorities in LDCs: An Application to Uganda.* New York: United Nations Department of Economic & Social Affairs.

Koop, S. H. A., Koetsier, L., Doornhof, A., Reinstra, O., Van Leeuwen, C. J., Brouwer, S., Dieperink, C. and Driessen, P. P. J. (2017). Assessing the governance capacity of cities to address challenges of water, waste, and climate change. *Water Resources Management, 31*(11), 3427–3443.

Lankao, P. and Gnatz, D. (2015). *Do cities have the institutional capacity to address climate change?* UGEC Viewpoints. Retrieved 23 November, 2020, from https://ugecviewpoints.wordpress.com/2015/04/07/do-cities-have-the-institutional-capacity-to-address-climate-change/

Leck, H. and Roberts, D. (2015). What lies beneath: Understanding the invisible aspects of municipal climate change governance. *Current Opinion in Environmental Sustainability, 13*, 61–67.

Lee, Y., Paredes, J. R. and Lee, S. H. (2012). *Smart Grid and Its Application in Sustainable Cities.* Washington DC: Inter-American Development Bank.

Lindfield, M. and Steinberg, F. (2012). *Green Cities.* Manila: Asian Development Bank.

Littler, C. R. (1978). Understanding Taylorism. *British Journal of Sociology, 29*(2), 185–202.

Martins, R. D. A. and Ferreira, L. D. C. (2011). Opportunities and constraints for local and subnational climate change policy in urban areas: Insights from diverse contexts. *International Journal of Global Environmental Issues, 11*(1), 37–53.

Maunsell Australian Pvt Ltd. (2008). *Impact of Climate Change on Australia's Telecommunications Infrastructure*. Garnaut Climate Change Review Report. Cambridge, UK: Cambridge University Press.

Meerow, S. and Newell, J. P. (2015). Resilience and complexity: A bibliometric review and prospects for industrial ecology. *Journal of Industrial Ecology*, *19*(2), 236–251.

Menes, R. (2017). *From Romania to Costa Rica – the persuasive power of green mortgages*. International Finance Corporation. Retrieved 23 November, 2020, from https://www.edgebuildings.com/from-romania-to-costa-rica-the-persuasive-power-of-green-mortgages/

Menes, R. (2019a). *Awareness builds for EDGE in India*. International Finance Corporation. Retrieved November 23, 2020, from https://www.edgebuildings.com/green-building-through-green-bonds-in-colombia-story/

Menes, R. (2019b). *Green building through green bonds in Colombia*. International Finance Corporation. Retrieved 23 November, 2020, from https://www.edgebuildings.com/green-building-through-green-bonds-in-colombia-story/

Met Office. (2020). *About UKCIP18*. Retrieved 23 November, 2020, from https://www.metoffice.gov.uk/research/approach/collaboration/ukcp/about

Michel-Kerjan, E., Hochrainer-Stigler, S., Kunreuther, H., Linnerooth-Bayer, J., Mechler, R., Muir-Wood, R., Ranger, N., Vaziri, P. and Young, M. (2013). Catastrophe risk models for evaluating disaster risk reduction investments in developing countries. *Risk Analysis*, *33*(6), 984–999.

Mitchell, J. F., Lowe, J., Wood, R. A. and Vellinga, M. (2006). Extreme events due to human-induced climate change. *Philosophical Transactions of the Royal Society A: Mathematical, Physical and Engineering Sciences*, *364*(1845), 2117–2133.

Ministry of Urban Development. (2015). *Smart Cities: Mission Statement and Guidelines*. Ministry of Urban Development, Government of India.

Modarres, M. (2008). Probabilistic risk assessment. In K. B. Misra (Ed.), *Handbook of Performability Engineering* (pp. 699–718). London: Springer.

Monahan, G. (2008). *Enterprise Risk Management: A Methodology for Achieving Strategic Objectives*, Vol. 20. Hoboken, NJ: John Wiley & Sons.

Mukhopadhyaya, A. and Jayal, N. (2000). *Decentralisation in India*. New Delhi: United Nations Development Program India.

Nafday, A. M. (2009). Strategies for managing the consequences of black swan events. *Leadership and Management in Engineering*, *9*(4), 191–197.

National Oceanic and Atmospheric Administration. (2013). *What Will Adaptation Cost? An Economic Framework for Coastal Community Infrastructure. Final Report.* Charleston, SC: National Oceanic and Atmospheric Administration Coastal Services Center.

National Oceanic and Atmospheric Administration. (2019). *What is storm surge?* Retrieved 23 November, 2020, from https://oceanservice.noaa.gov/facts/stormsurge-stormtide.html

New York Times. (2018). *On climate, Gov. Murphy brings a new voice to New Jersey*. Retrieved 23 November, 2020, from https://www.nytimes.com/2018/01/29/opinion/jersey-murphy-climate-change.html

NGWA. (2015). *Groundwater supply and use*. The Groundwater Association. Retrieved 23 November, 2020, from https://wellowner.org/groundwater/groundwater-supply-use/

Nightingale, A. J., Eriksen, S., Taylor, M., Forsyth, T., Pelling, M., Newsham, A. T., Boyd, E., Brown, K., Harvey, B., Jones, L., Bezner Kerr, R., Mehta, L., Naess, L. O., Ockwell, D., Scoones, I., Tanner, T. and Whitfield, S. (2020). Beyond technical fixes: Climate solutions and the great derangement. *Climate and Development*, *12*(4), 343–352.

NYC Mayor's Office of Resiliency. (2020). *New York City Panel on Climate Change*. Retrieved 23 November, 2020, from https://www1.nyc.gov/site/orr/challenges/nyc-panel-on-climate-change.page

Onwutuebe, C. J. (2019). Patriarchy and women vulnerability to adverse climate change in Nigeria. *Sage Open*, *9*(1), Article 2158244019825914.

Overseas Development Institute. (2015). *Measuring Urban Climate Change Resilience*. London: Overseas Development Institute.

Pelling, M., High, C., Dearing, J. and Smith, D. (2008). Shadow spaces for social learning: A relational understanding of adaptive capacity to climate change within organisations. *Environment and Planning A*, *40*(4), 867–884.

Ponzi, D. and Iwasaki, H. (2014). *Climate Proofing ADB Investment in the Transport Sector: Initial Experience*. Manila: Asian Development Bank.

Prichard, B. (2013). The North Sea surge and east coast floods of 1953. *Weather*, *68*(2), 31–36.

Rahmstorf, S. (2017). Rising hazard of storm-surge flooding. *Proceedings of the National Academy of Sciences*, *114*(45), 11806–11808.

Ramalingam, B., Jones, H., Reba, T. and Young, J. (2008). *Exploring the Science of Complexity: Ideas and Implications for Development and*

Humanitarian Efforts, Vol. 285. London: Overseas Development Institute.

Ramalingom, R. and Kurup, K. N. (1991). Plan transfers to states: Revised Gadgil Formula: An analysis. *Economic and Political Weekly*, 501–506.

Reed, S., Friend, R., Toan, V. C., Thinphanga, P., Sutarto, R. and Singh, D. (2013). "Shared learning" for building urban climate resilience – experiences from Asian cities. *Environment and Urbanization*, *25*(2), 393–412.

Reeves, S., Winter, M., Leal, D. and Hewitt, A. (2019). *Roads: An Industry Guide to Enhancing Resilience*. London: Resilience Shift, UK.

Revi, A., Satterthwaite, D. E., Aragón-Durand, F., Corfee-Morlot, J., Kiunsi, R. B. R., Pelling, M., Roberts, D. C. and Solecki, W. (2014). Urban areas. In C. B. Field, V. R. Barros, D. J. Dokken, K. J. Mach, M. D. Mastrandrea, T. E. Bilir, M. Chatterjee, K. L. Ebi, Y. O. Estrada, R. C. Genova, B. Girma, E. S. Kissel, A. N. Levy, S. MacCracken, P. R. Mastrandrea and L. L. White (Eds), *Climate Change 2014: Impacts, Adaptation, and Vulnerability. Part A: Global and Sectoral Aspects. Contribution of Working Group II to the Fifth Assessment Report of the Intergovernmental Panel on Climate Change* (pp. 535–612). Cambridge: Cambridge University Press.

Roberts, N. C. and King, P. J. (1991). Policy entrepreneurs: Their activity structure and function in the policy process. *Journal of Public Administration Research and Theory*, *1*(2), 147–175.

Ruth, M. and Coelho, D. (2007). Understanding and managing the complexity of urban systems under climate change. *Climate Policy*, *7*(4), 317–336.

Sanders, T. I. (2008). Complex systems thinking and new urbanism. In T. Haas (Ed.), *New Urbanism and Beyond: Designing Cities for the Future* (pp. 275–279). New York: Rizzoli.

Shah, T., Mavalankar, D., Ganguly, P., Dutta, P., Tiwari, R., Jaiswal, A., Knowlton, K., Connolly, M., Kaur, N., Deol, B., Hess, J. and Sheffield, P. (2018). *Innovative Heat Wave Early Warning System and Action Plan in Ahmedabad, India*. Geneva: World Health Organisation/World Meteorological Organisation.

Shakya, C., Cooke, K., Gupta, N., Bull, Z. and Greene, S. (2018). *Building Institutional Capacity for Enhancing Resilience to Climate Change: An Operational Framework and Insights From Practice*. Action on Climate Today Learning Paper. New Delhi: ACT.

Song, L. K. (2016). Planning with urban informality: A case for inclusion, co-production and reiteration. *International Development Planning Review*, *38*(4), 359–381.

State of New Jersey. (2019). *Governor Murphy signs executive order to establish statewide climate change resilience strategy*. Retrieved 23

November, 2020, from https://www.nj.gov/governor/news/news/562019/approved/20191029a.shtml

Stephens, J. C., Wilson, E. J., Peterson, T. R. and Meadowcroft, J. (2013). Getting smart? Climate change and the electric grid. *Challenges*, *4*(2), 201–216.

Tanner, T., Zaman, R. U., Acharya, S., Gogoi, E. and Bahadur, A. (2019). Influencing resilience: The role of policy entrepreneurs in mainstreaming climate adaptation. *Disasters*, *43*(S3), S388–S411.

The Energy Resources Institute. (2015). *State-wise training programs*. Retrieved 23 November, 2020, from https://www.teriin.org/projects/apn/odisha.php

Tonmoy, F. N., Rissik, D. and Palutikof, J. P. (2019). A three-tier risk assessment process for climate change adaptation at a local scale. *Climatic Change*, *153*(4), 539–557.

Tyler, S. and Moench, M. (2012). A framework for urban climate resilience. *Climate and Development*, *4*(4), 311–326.

UGE. (2020). *UGE partners with EDC to help 16 small businesses in NYC on the flood zone install microgrid systems for energy resiliency*. Retrieved 23 November, 2020, from https://www.ugei.com/rise-case-study

UN-Habitat. (2017). *Trends in Urban Resilience*. Nairobi: UN-Habitat.

United Nations Framework Convention on Climate Change. (2019a). *Annual Technical Progress Report of the Paris Committee on Capacity-Building*. Bonn: United Nations Framework Convention on Climate Change.

United Nations Framework Convention on Climate Change. (2019b). *Differentiated Impacts of Climate Change on Women and Men; the Integration of Gender Considerations in Climate Policies, Plans and Actions; and Progress in Enhancing Gender Balance in National Climate Delegations*. Bonn: United Nations Framework Convention on Climate Change.

United Nations Framework Convention on Climate Change. (2020). *What do adaptation to climate change and climate resilience mean?* Retrieved 23 November, 2020, from https://unfccc.int/topics/adaptation-and-resilience/the-big-picture/what-do-adaptation-to-climate-change-and-climate-resilience-mean

United Nations General Assembly. (2015). *Transforming Our World: The 2030 Agenda for Sustainable Development*. New York: United Nations General Assembly.

United Nations General Assembly. (2016). *Report of the Open-Ended Intergovernmental Expert Working Group on Indicators and Terminology Relating to Disaster Risk Reduction*. New York: United Nations General Assembly.

United Nations Office for Disaster Risk Reduction. (2017). *Disaster Resilience Score Card for Cities*. Geneva: UN Office for Disaster Risk Reduction.

United States Department of Energy. (2020). *The smart grid*. Retrieved 23 November, 2020, from https://www.smartgrid.gov/the_smart_grid/smart_grid.html

United States Environmental Protection Agency. (2015). *Climate impacts on energy*. Retrieved 23 November, 2020, from https://19january2017snapshot.epa.gov/climate-impacts/climate-impacts-energy_.html

United States Geological Survey. (2020). *California drought*. Retrieved 23 November, 2020, from https://ca.water.usgs.gov/california-drought/california-drought-comparisons.html

Vandeveer, S. D. and Dabelko, G. D. (2001). It's capacity, stupid: International assistance and national implementation. *Global Environmental Politics*, *1*(2), 18–29.

Vatsa, K. (2019). *Equipping states for disaster*. United Nations Development Programme India. Retrieved 23 November, 2020, from https://www.in.undp.org/content/india/en/home/blog/equipping-states-for-disaster.html

Venema, H. and Temmer, I. (2017). *Building a climate-resilient city: Electricity and information and communication technology infrastructure*. International Institute for Sustainable Development and the University of Winnipeg.

Verougstraete, M. and Enders, I. (2014). *Efficiency Gains: The Case of Water Services in Manila*. Bangkok: United Nations Economic and Social Commission for Asia and the Pacific.

Weber, M. (2009). *The Theory of Social and Economic Organization*. New York: Simon and Schuster.

Welp, Y., Urgell, F. and Aibar, E. (2007). From bureaucratic administration to network administration? An empirical study on e-government focus on Catalonia. *Public Organization Review*, *7*(4), 299–316.

Willems, S. and Baumert, K. (2003). *Institutional Capacity and Climate Actions*. Paris: Organisation for Economic Co-operation and Development.

World Bank. (2014). *Can Tho, Vietnam: Enhancing Urban Resilience*. Washington DC: The World Bank Group.

World Health Organization. (2015). *Operational Framework for Building Climate Resilient Health Systems*. Geneva: World Health Organization.

World Health Organization. (2019). *Water shortage in the Philippines threatens sustainable development and health*. Retrieved 23 November, 2020, from https://www.who.int/philippines/news/feature-stories/detail/water-shortage-in-the-philippines-threatens-sustainable-development-and-health

Chapter 6

Urban resilience finance: From exogenous reliance to endogenous reliability

As areas with high risks and significant vulnerabilities, urban areas face very significant financing needs to invest in building resilience. Yet most towns and cities face considerable barriers in accessing finance and lack established mechanisms to determine who should pay for investments. As a result, urban areas can be acting as "involuntary pioneers" in the development and deployment of financing mechanisms (Plastrik et al. 2019: 4). While an experimentation phase is well under way and lessons are being learned and shared, much investment has been conceived, developed and sourced externally to the urban sphere. This has come in particular from international public finance and philanthropic funding, which is able to provide pioneer finance but may find it harder to generate internal ownership of agendas and structural changes, integrate resilience into mainstream actions or secure longer-term sustainability.

This chapter calls for a pivot away from this funding reliance to unlock greater levels of endogenous finance that is conceived and controlled by governments, businesses and households. We argue that this pivot requires greater attention to supportive enabling environments to generate endogenous finance, upscaling of innovative financing tools, recognition of informality and equity, and a focus on developing business cases that recognise and capture the multiple benefit streams of resilience building in urban areas.

6.1 FINANCING URBAN RESILIENCE: NEEDS AND OPTIONS

6.1.1 Financing needs

As highlighted throughout the book, climate-related risks are growing in urban environments in most countries globally. These risks are more acute in cities of the Global South, where the concentration of assets is combined with dense and growing populations that experience large inequalities, large informal settlements and inadequate service provision and safety nets. Specific vulnerabilities are exacerbated by the location of many of these cities in low-lying and coastal locations exposed to floods, storms and heatwaves.

In the context of these risks and vulnerabilities, it is perhaps unsurprising that financing needs for climate change resilience are considerable. Estimates of the annual global costs of adapting to climate change range from USD 140 billion to USD 300 billion by 2030 and from USD 280 billion to USD 500 billion by 2050 (UNEP 2018). These figures are likely to underestimate the reality as biodiversity and ecosystem services are omitted. While there is growing work to identify these adaptation gaps in terms of investment (e.g. through the Adaptation Gap reports of the United Nations Environment Programme and partners; see UNEP 2018), little has been done to quantify these gaps in an urban context (Chen et al. 2016).

Much of the focus around financing urban climate change resilience has been on infrastructure development (CCFLA 2015; World Bank 2018). This reflects the dominance of infrastructure in the scale of investments needed, with physical changes generally far costlier than social or institutional interventions. It also reflects the relative ease with which estimates of financing needs can be made given the knowledge on existing infrastructure, retrofitting and future investment projections. Urban infrastructure resilience estimates are often calculated in terms of the additional costs due to an incremental shift in impacts due to climate change, generally related to existing urban footprints and infrastructure. This may underestimate costs as it does not pick up the pre-existing impacts of historical climate variability or the growth of exposed assets in the future. Cost is further underestimated in the Global South, where the under-provision of basic services means they have currently less services and infrastructure than they require (Satterthwaite et al. 2020). Building resilience therefore requires additional financing to address this deficit as well as for retrofitting existing infrastructure.

Nevertheless, urban areas dominate global adaptation and resilience costings, principally because they contain more high-cost infrastructure and have a higher concentration of the population. Under a low-carbon future climate change scenario, up to 2030 roughly 70% of the required global infrastructure investment is needed for urban areas; between 9% and 27% per annum (USD 0.4 trillion to USD 1.1 trillion) of additional investment is needed to make these investments low-emission and climate-resilient (CCFLA 2015; Alexander et al. 2019). These figures highlight how the financing of resilience is situated within a wider and far greater investment challenge in many cities of the world.

While data sets on resilience spending are still in their infancy, there is evidence that spending on urban infrastructure is growing. High-profile international programmes such as the ICLEI Resilient Cities Congress series, the United Nations International Strategy for Disaster Risk Reduction's Making Cities Resilient campaign and the 100 Resilient Cities initiative have helped to gain high-level political support. Within cities, evidence of spending in the urban "adaptation economy" shows significant flows, albeit still representing a small proportion of overall spending (see Figure 6.1).

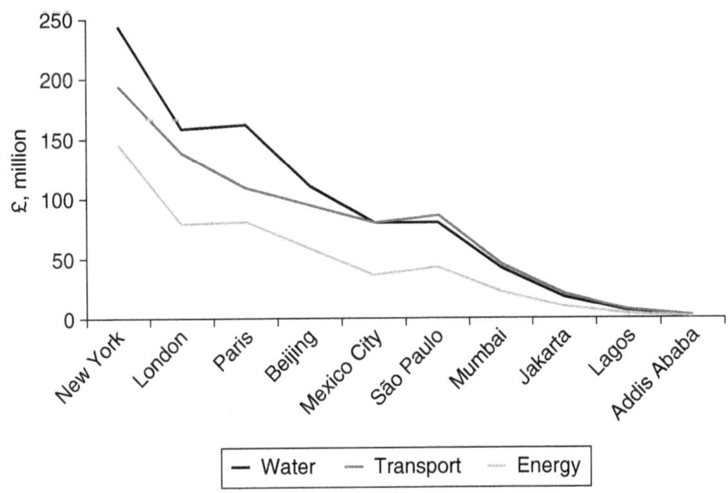

FIGURE 6.1 Total climate adaptation spending by urban infrastructure sector in ten megacities, 2014–2015. Source: World Bank (2018: 27).

While one way of framing the shortfalls in finance has been in relation to an adaptation or resilience "gap", these financing requirements can usefully be viewed as an opportunity to pursue more equitable development pathways. We argue in this chapter that the growing understanding and appetite for urban resilience is reflected not only in urban plans and public sector spending, but also in businesses in both the formal and informal economies. Influencing the form and nature of these emerging financing processes provides an opportunity to create societies and natural and built environments that are more equitable and resilient. The threat of climate change and the development of climate-resilient public infrastructure both serve to reduce investment risk and catalyse private investments (Schwarze et al. 2018).

6.1.2 Existing options for financing urban resilience

In order to meet resilience investment needs, a wide range of policy mechanisms and instruments are available, most of them common to wider urban financing. We briefly highlight some of the dominant modes, highlighting their main application here (as summarised in Table 6.1). We then outline the key challenges facing urban resilience financing in Section 6.2.

Many resilience investments have the characteristics of public goods, meaning there is a tendency to focus on funding from municipal sources. Municipal sources are important as recent processes of decentralisation have given far greater power to urban bodies in many emerging economies, and this is combined with the challenges of urban service delivery in the context of rapid and unplanned urban population growth (Gorelick 2018). Such finance may be generated locally or transferred from central government tax revenues to local authorities. Taxes collected from general sources such as income, property or vehicle use could be used to fund resilience investments, but these revenues are often already under severe pressure to be allocated to other activities. More targeted location-specific charges or taxes with links to resilience may therefore be more effective. These may be linked, for example, to the use of resilient infrastructure or resilience services, such as bridges, wetlands or forecasting.

We argue later in this chapter that catalysing endogenous public finance is crucial to successful future investment in urban resilience.

TABLE 6.1 Options for financing urban resilience

Actor	Source of finance	Examples of how these can be applied to urban resilience
Municipal	Central budget allocations, property taxes, user charges, tourism taxes/fees, impact fees, betterment levies, land value capture, vehicle taxes, licences/registrations	• Maintain infrastructure assets and retrofit small capital investments • Set rates to cover differential costs of adaptation expenditures • Cover cost of incremental extension or upgrading of infrastructure and services • Cover personnel costs for research and advisory services
Intergovernmental	Earmarked grants, conditional grants, shared taxes, programmatic transfers, revenue sharing	• Tie grants to programmes and give cities flexibility to decide how to spend funds • Make grant disbursement conditional on reforms to local public administration and adaptation policy, programmes and expenditures
Multilateral	International donor funds, official development assistance, humanitarian aid, technical and capacity support	• Provide concessional loans through national and/or regional governments for local infrastructure and service delivery improvements • Provide direct technical assistance to city governments to mainstream climate adaptation in urban planning and/or project preparation • Lend support during and after climate/disaster emergencies for rebuilding and rehabilitation
Private	Municipal bonds, loans, private investments, insurance/reinsurance, individual private capital, pooled finances from communities	• Support private investment or public–private partnerships in local adaptation services • Provide long-term finance for green infrastructure • Generate income for larger-scale investments in infrastructure and building adaptation • Fund household and/or community investments in adaptation measures such as housing upgrades to reduce flood and storm risk

(continued)

TABLE 6.1 Cont.

Actor	Source of finance	Examples of how these can be applied to urban resilience
Non-profit	Philanthropy, individual donations, microfinance	• Solicit donations for targeted urban adaptation capacity programmes for local officials and communities • Cover costs of future welfare losses from weather or other climate-related events • Build local capacity across urban stakeholders to adapt to climate change

Source: Adapted from Chu et al. (2019); Cook and Chu (2018)

Some city authorities are arguably acting faster than their national government counterparts, including through improving the institutional basis for financing through the mainstreaming of resilience concerns into wider sectoral plans and policies. However, as outlined in Section 6.2, many of them remain constrained in terms of their fiscal capacity to raise revenues.

We contend in this chapter that the global urban resilience narrative has been dominated by the financing and conceptual leadership of actors external to the urban areas in question. Sources of finance for urban resilience are increasingly found in intergovernmental agencies and multilateral banks. A review of multilateral climate funds between 2010 and 2014 by Barnard (2015) found 47 of over 700 projects had explicit urban climate objectives, totalling just over 11% of the USD 7.28 billion approved by multilateral climate funds for all projects in the same period. Urban agendas have been significantly expanded, particularly since the 2008 global financial crisis, playing a countercyclical role as financing from commercial banks dried up.

Urban resilience now features in all the major lenders' strategic goals and portfolios, with dedicated multilateral climate funds beginning to reflect urban areas as a priority for support. The Asian Development Bank's Urban Climate Change Resilience Trust Fund is designed to integrate resilience concerns into its wider financing modalities in medium-sized cities. The Green Climate Fund portfolio has over USD 6 billion of committed spending and includes both infrastructure and cities in its results areas. Although initially slow to focus on urban-specific projects, these are now starting to emerge in the portfolio. Other significant dedicated funding sources relevant to urban resilience include the Climate Investment Funds, the Global Facility for Disaster

Reduction and Recovery, the GEF, the Sustainable Development Goals Fund, the Millennium Development Goals Achievement Fund, and United Nations Framework Convention on Climate Change funds including the Adaptation Fund and Least Developed Countries Fund. These multilateral sources employ a mix of financing mechanisms including grants and subsidies, concessional rate loans, and de-risking instruments such as guarantees and public equity co-investments.

Philanthropic finance has also been a considerable catalyst, with the Rockefeller Foundation in particular pushing urban resilience agendas through its Asian Cities Climate Change Resilience Network, which invested USD 59 million as well as leveraging further financing across Asian cities. In parallel, the foundation's 100 Resilient Cities initiative (and its legacy, the Global Resilient Cities Network) has worked with cities around the world to develop and implement resilience strategies and share knowledge and resources. As well as funding implementation, these programmes have helped to develop awareness and establish global networks on urban resilience, putting the issue on urban development agendas and emphasising the needs of vulnerable residents.

Acknowledging that financing needs hugely outstrip the capacities of the public sector, there is increasing attention on how to draw in the significant private sector flows to support urban resilience. The main existing flow is either through direct investment to protect business assets or through taking on general or project-specific equity or debt. Equity and debt-based finance are already a common part of urban investment, particularly in infrastructure projects, but greater efforts are needed to highlight their resilience dimensions and selling points. Insurance mechanisms also provide opportunities for supporting resilience, though private risk or catastrophe insurance and via national or regional parametric risk facilities (World Bank 2018). Crucially, insurance needs to go beyond simply providing finance for the repair or replacement of damaged assets to focus on incentivising future risk management and resilience.

There are growing examples of companies taking action to adapt their own operations to manage climate risks, and this is in relation to both infrastructure and soft management systems such as logistics or supply chain flexibility. This can enhance individual competitiveness, but businesses may also invest in wider urban resilience to support the buildings, communications, transport systems, human capital and potential markets on which they depend (Crawford and Seidel 2013; Schwarze et al.

2018). Businesses can offer knowledge and technical skills that contribute to wider efforts needed to tackle urban resilience, formally through public–private partnerships (that also provide links to financing) or through knowledge-sharing networks.

6.2 CHALLENGES FOR URBAN RESILIENCE FINANCING

The previous section suggested that the financing needs for resilience are high despite the range of potential sources of finance. We argue here that the untapped potential for resilience-related investment can be explained by a set of challenges related to knowledge, finance, institutions and politics. The emergence of the COVID-19 pandemic has created further challenges to the prioritisation of investments in resilience, but it has simultaneously created potential conditions for synergies.

6.2.1 Capacity gaps and financial constraints in creating "bankable" resilience projects

Many cities in the Global South lack the knowledge and capacity to effectively exploit existing sources of finance or develop new channels. This is often framed in relation to low awareness of international funding options, limited capacity to develop project proposals in the formats required or lack of counterpart funding from city authorities (Beltran 2012). Information and data that could support project development for resilience building interventions are often unavailable.

Creating bankable projects around resilience faces barriers due to the nature of resilience itself, information needs and the levels of knowledge and experience around such investments. Bankability requires not just profitability but also an ability to quantify future outcomes, including social or environmental benefits, as well as clear implementation and evaluation mechanisms (ICLEI 2019). Yet the costs of disaster events or gradual climate-related impacts are not always clearly known or understood. Both climate-related impacts and resilience building options may have a wide range of costs and benefits distributed across different stakeholder groups, geographical spaces and timescales. This is often compounded by the paucity of historical data on hazards or disaster impacts, which can frustrate attempts to quantify benefits.

Decision makers hence underinvest in resilience building projects because although the costs are visible and immediate, the benefits are unclear, uncertain and distant. Investment decisions often fail to incentivise resilience building because the resulting benefit streams are undervalued (Surminski and Tanner 2016). Where the benefits of financing resilience are not clear, it is far harder to make the case for allocating scarce resources. Resourcing is commonly stretched, with existing deficits limiting the ability to deliver basic services to the urban population. Many resilience investments are also, by their nature, publicly rather than privately oriented, meaning their financial returns may not be sufficient for private investors and commercial loan sources. At the same time, capital-intensive projects that do not include ongoing charges for end users may be seen as financially weak (Beltran 2012).

Fundamentally, cities often lack funds to cover costs of project preparation and indeed may be poorly prepared for raising finance generally (CCLFA 2015). Many municipalities, especially in poorer countries, have limited resources and limited capacity to collect local tax revenues. Local government resources for investment projects are typically limited. Indonesian local governments, for example, spend around 75% of their financial allocation on maintenance costs of public staff and infrastructure, leaving only 25% for investment programmes (Beltran 2012). Many cities have either poor creditworthiness or lack a credit rating entirely, reflecting a limited track record and underdeveloped capital markets for local government borrowing. Moreover, private sector investors may be geared to national rather than city-level investment opportunities.

These factors combine to limit the capacity for financing of climate change adaptation and resilience in cities. A C40 survey in 2018 found an extremely low number of climate adaptation projects, which they attributed to potential difficulties in conceptualising viable projects, particularly in terms of identifying and measuring the benefits related to climate change adaptation (C40 2018).

6.2.2 Institutional and political challenges

Even where there is technical capacity, political will and an operational spending plan for investment in urban resilience, institutional factors may constrain access to finance. One factor is that urban governments have insufficient judicial power to create revenues through taxes, levies or other mechanisms. Similarly, options for borrowing at subnational level, including issuing municipal bonds, may be constrained by the limited legal powers and poor credit ratings of urban governments. National governments create the overall operational environment for cities but would also ultimately bear the unpaid debt of urban authorities, and so they have a fiduciary responsibility to influence their fundraising powers and risk profiles.

National governments also influence official development assistance and the types of project that are prioritised for these funds, with few lending directly to city authorities. Clients are essentially sovereign countries – not cities – so national governments lead on developing strategies and determine the areas and projects that are ultimately financed. Reviewing financing options for climate change in urban areas, Alexander and colleagues (2019: 16) conclude that the ownership, mandates, operational practices and institutional arrangements of the multilateral development banks are in reality "not geared to respond to the climate change investment needs of cities".

Resilience issues and mandates are often spread across ministries, functions and scales, making them difficult to track and leading to accountability issues. The cross-sector and multisystems nature of resilience, with often overlapping or unclear roles and responsibilities of national and city governments and agencies, thereby frustrates implementation (Bahadur and Tanner 2014a). Such blurred boundaries are evident in many cities, including in Indonesia across agencies for waste, the environment agency and public works agency (Beltran 2012). The 100 Resilience Cities initiative attempted to address this concern by providing funding for chief resilience officers in each city, but the abilities of these officers as coordinators were constrained where the postholder did not also have power to control city planning or spending decisions (Martín and McTarnaghan 2018).

There is growing recognition of the role of politics in mediating urban risk and resilience (Bahadur and Tanner 2014b; Allen et al. 2017). At a wider political scale, there may be differences in political priorities between national governments and city administrations. At the urban scale, frequent political changes can prevent action as new administrations do not see resilience as a priority spending area or they seek short-term visible impacts. By contrast, resilience suffers from a lack of salience with citizens, as the benefits are hard to perceive. This is compounded by the perception of decision makers that resilience investments yield benefits predominantly over the long term or only in the event of disasters (Tanner et al. 2015). There are also political biases favouring crisis response over anticipatory investment due to the political visibility of relief efforts and the related reinforcement of patron–client networks (Wilkinson 2012). In addition, the promise of international aid in the event of a disaster event distorts political incentives to invest. The redirection of funding following the COVID-19 pandemic demonstrates how emergent crises can divert attention from more anticipatory action.

There are, therefore, political challenges when it comes to orienting finance to consider anticipation and equity in resilience investments. In this regard, we echo both the criticism that resilience lacks normative dimensions and the calls for a more critical analysis of rights, power and justice in planning and implementing urban resilience initiatives. The latter covers procedural dimensions (who is involved), distributional dimensions (the likely outcomes of capacity building), rights dimensions (who is entitled to set the agenda) and responsibility dimensions (who has the capacity to contribute effectively) (Archer and Dodman 2015; see also Ziervogel et al. 2017; Matin et al. 2018).

While international financing, especially through official development assistance (aid), may claim to target poor and vulnerable citizens, even this does not guarantee that it can be accessed by these groups or that others will work in their interests. As Ayers (2009: 226–7) notes, “raising funds for adaptation at the international level is not enough as long as barriers to accessing and using these funds for the most vulnerable groups in the most vulnerable countries exist”. Analysis of urban resilience capacity development activities in Bandar Lampung, Indonesia, and Hat Yai, Thailand, bear these issues out, demonstrating the gaps in targeting, both in reaching the poorest and those most vulnerable to climate change impacts and in influencing those with most power over policymaking (Archer and Dodman 2015).

6.3 PIVOTS TO UNLOCK FINANCE FOR URBAN RESILIENCE

The challenges outlined above serve to constrain existing finance for urban resilience. We argue in the remainder of this chapter that a set of critical pivots are required in order to re-orient financing in the future to develop endogenously led mechanisms and provide a more socially inclusive approach. Existing financing approaches will disproportionately benefit large and hub cities that are already creditworthy, so for most other urban centres, innovative financial and collaborative approaches are needed to develop bankable projects, markets and mobilise financing (IFC 2018).

Pivots include enhancing the processes to support enabling environments for finance, including legal and institutional regimes, that can improve creditworthiness, as well as enhancing the technical capacity to develop projects that can more readily attract investment. With the support of an improved enabling environment, we suggest a pivot towards greater engagement with businesses and scaling up of innovations for financing urban resilience including bonds, insurance and land value capture. Finally, we argue that narratives and business cases can benefit significantly from a re-orienting of financing logics towards the multiple benefits and dividends that urban resilience investments can deliver. This pivot is framed by an orientation towards equity and justice that can be supported by engagement with bottom-up approaches to autonomous innovation going forward.

6.3.1 Supporting enabling environments for innovative and endogenous finance

When considering support for climate resilience, there are compelling reasons to target projects with visible short-term impacts with limited available financing. International public finance sources can be under pressure from their stakeholders (the governments and citizens of donor countries) to account for delivery of tangible benefits. In the urban resilience realm, these sources have played an important role in stimulating early action, demonstration and knowledge sharing. However, they also suffer from limitations, including sustainability of intervention activities beyond the project lifetime; bias towards hard infrastructure and technical resilience solutions (Nightingale et al. 2020); difficulty in generating more structural changes (e.g. in relation to legal regimes) over the timescales in which projects typically operate; and being led by external donor priorities and narratives, conditionalities and modalities.

An essential pivot for urban resilience is therefore to support the development of legal, institutional and policy regimes to overcome challenges outlined in Section 6.2. As noted in the Cities Climate Finance Leadership Alliance report (2015: 5), "[t]he challenge is not simply to increase the amount of money in the pipeline, but also to create an enabling environment that encourages existing and new financing to flow from a broad spectrum of sources". Such support is particularly critical in smaller cities that require policies that can underpin their creditworthiness, develop climate finance ecosystems and integrate resilience considerations into development policies and planning (IFC 2018).

Banks also provide technical assistance to support technological, institutional and organisational solutions to urban development. Their support can also add credibility and transparency to projects as a result of the due diligence processes required, which can help maintain credit ratings and crowd in other finance (Future Cities Catapult 2014). There is already evidence of this shift in the enabling support in the forms of technical assistance offered by multilateral development banks (see Figure 6.2). For example, the World Bank hosts a City Resilience Program providing technical assistance to work with city authorities to develop project proposals that can attract finance from international public sources and the private sector. In addition, its City Creditworthiness Initiative provides technical assistance and training to strengthen financial performance and develop legal, institutional and policy frameworks. The Asian Development Bank's Urban Climate Change Resilience Trust Fund has also strengthened its support to environments that enable urban resilience, including for the institutional capacities needed to identify, plan, invest in, and respond to climate change and disaster-related risks.

However, such initiatives need to move beyond those cities with existing enabling environments and fledgling innovation in resilience finance. What is important is that these initiatives do not seek easy wins from working with global megacities and work instead with smaller, often poorer municipalities with high levels of vulnerability. International public finance also has a role to play in developing endogenous capacity to deliver financing modalities. The concerns noted here echo those expressed in the 100 Resilient Cities process over the dominance of global consultancies and the "fear of colonization by global players that might marginalize local expertise and practices" (Webber et al. 2020: 354).

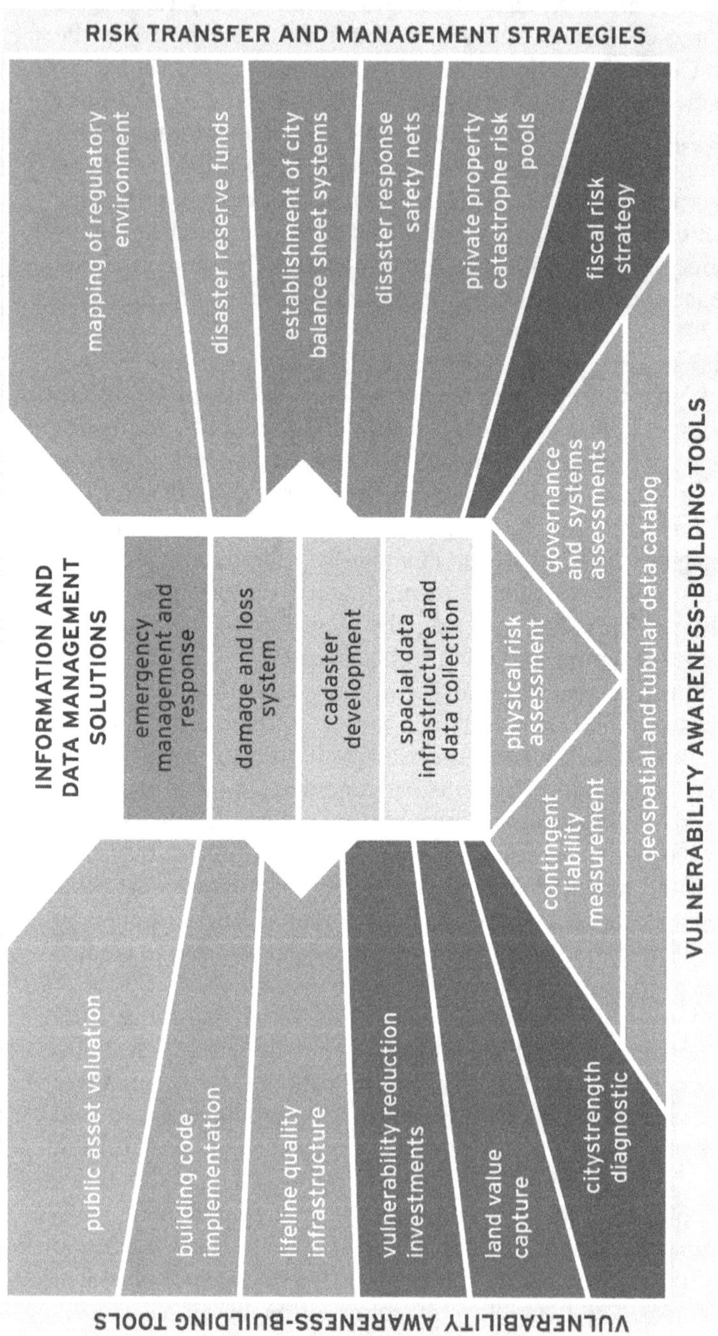

FIGURE 6.2 Potential technical assistance options to support urban resilience finance. Source: World Bank (2015: 76).

Targeted support to develop enabling environments can enhance creditworthiness and help to reduce dependence on predominantly grant-based finance that is tied to donor interests and move towards more demand-responsive sources directed by urban bodies. Crucially, this needs to engage private investors and businesses who could invest autonomously in enhancing resilience in their own and neighbouring operations as well as partnering with city authorities to do so. Kennan et al. (2019) refer to this as a shift "from funding to financing", where funding refers to direct payments by public or private actors for building resilience, while financing refers to market-based mechanisms.

At the same time, far more effort is needed to reflect the financing needs of urban residents, which are often missed in formal mechanisms. As highlighted in Chapter 4, the informal nature of significant parts of city economies means that municipal government financing may not influence the systems within which large proportions of the urban population live. While government-led initiatives can be more targeted to the informal economy and informal settlements, there is growing awareness that they may not reach all citizens and that community-led resilience interventions may be better able to adapt to informality, serve local needs and deliver equitable outcomes (Satterthwaite et al. 2020).

Weru and colleagues (2018) describe the funding and financial services provided by the Akiba Mashinani Trust for the Kenyan Homeless People's Federation (Muungano wa Wanavijiji); this reflects an emerging approach to promoting resilience to shocks and increasing capacity to generate income. The trust is a federation of autonomous savings groups and has over 60,000 members in informal settlements across Kenya, facilitating not only saving but also demonstrable capacity for loan repayment that opens access to other finance mechanisms. The funds provide products to informal settlers, including community project loans through savings groups that in turn provide affordable finance for social housing, sanitation and basic infrastructure. Unlike most of the financial sector in Kenya, these mechanisms specifically target the informal sector as well as serving to strengthen social capital among those living in informal settlements.

6.3.2 Private sector resilience

A crucial part of the pivot towards endogenously driven finance for resilience is engagement with companies in urban areas. This includes large and small companies, local and international companies and those in the formal and informal sectors. While innovative financing mechanisms are often geared to enhancing the role of the private sector through investment finance for wider resilience building activities, there is a crucial role for business in improving resilience within their own operations. The economic impacts of climate change in cities are borne primarily by businesses – directly, indirectly and via the wider economy. Private sector actions and partnerships therefore play a major role in finance to build resilience.

A range of incentives are driving urban resilience investments within businesses rather than simply as lending for public investment (Bahadur et al. 2016; Schwarze et al. 2018). While protecting company assets and supply chains is of particular interest, incentives extend beyond economic benefits of limiting direct and indirect losses. Legal compliance increasingly drives investment in urban resilience, as exemplified in the 2016 European Union legislation requiring that EUR 3 trillion of pension funds must show how they will consider the impact of climate risks on their investment strategies (Acclimatise 2017). Social or environment responsibility concerns also provide an incentive, including the duty of care to employees as well as broader responsibility to protect social welfare. As noted in Section 6.3.6, there is also a growing acknowledgement that investments in resilience building can yield a range of co-benefits.

Businesses are increasingly aware of the impacts of climate change on their operations, both directly and indirectly. This has already prompted some companies, particularly larger and multinational corporations, to strengthen their resilience strategies, particularly through risk management of their supply chains (Crawford and Seidel 2013). Toyota were prompted to act after the 2011 floods in Thailand slowed production in eight different countries (Avory et al. 2015). The 2004 flooding in Dhaka, Bangladesh, decreased production in the garment factories that support the country's export-led growth. As well the floods affecting the factories directly, this event illustrated to factory owners that staff, who often live in the city's highly populated low-lying areas, were prevented from accessing their workplace, thus increasing production losses (Alam and Rabbani, 2007).

As companies become increasingly engaged in global value chains, there is greater incentive to protect physical assets in a wider range of geographical locations, with cities frequently acting as hubs. Such actions can safeguard an individual company's own interests, but can also support other, often smaller, companies in the value chain. Local chambers of commerce and/or business associations can also educate small businesses on climate change risks (see the Surat example later in this section). Resulting activities can include capacity building, identifying incentives and opportunities and sharing risk information to protect small businesses.

Building resilience in key urban utilities is often considered a public sector task, but the reality in many cities is that utilities such as energy, communications, transport and water systems are contracted out to private firms. As a result, there are opportunities to develop existing and future markets, such as supplying robust urban infrastructure, hard and soft technologies or risk information. Mutual aid agreements between municipal governments and firms can formalise forms of disaster support; this may be particularly important for those sectors working with more vulnerable people or at-risk urban locations (e.g. small-scale construction in informal settlements), as this work can lie outside formal regulatory systems, as highlighted in Chapter 4.

The limited availability of financing from public sector sources is likely to lead to growth in public–private partnerships for enhancing climate change resilience. These include contracting out the delivery of public sector services to the private sector as well as full-scale joint ventures. Local NGOs and civil society organisations can also support firms to deliver their corporate social responsibility goals; for instance, finance and insurance companies are increasingly developing risk transfer mechanisms for urban communities. Collaborative initiatives among companies may also develop, as seen in the Southern Gujarat Chamber of Commerce & Industry, which formed the Surat Climate Change Trust in India (discussed in Section 3.1.3). The trust provides a platform for public, private and civil society actors to come together to prioritise adaptation options, seek financial support and define the city resilience agenda to promote resilience among sectors and communities vulnerable to the impacts of climate change and urbanisation (Chu, 2016).

Municipal authorities can also use financial incentives to engage the private sector in resilience building. These include supportive changes to business, sales and property taxes; rebates to promote installation of design features that incorporate robustness, redundancy and flexibility, such as flood-proofing or information technology backup systems; subsidies, grants and soft loans; and financial aid following a disaster (Khattri et al. 2010; Jha et al. 2013). There may also be incentives based on fulfilling explicit requirements within partnership contracts, such as business continuity plans and insurance provisions, in hazard-prone locations with a history of disaster-related events. (Bahadur et al. 2016).

6.3.3 Innovative financing modalities: Green municipal bonds

Part of a pivot towards endogenous reliance lies in the further development of financing modalities that can overcome some of the existing challenges of urban resilience finance. One potential mechanism for raising urban resilience finance is municipal bonds issued by urban public authorities, which provide a means of raising debt-related finance at the city level. Historically, some municipal bonds have been used for sustainable urban infrastructure and services. The first emerging economy city to raise bonds, Johannesburg in South Africa has been lauded as a success story in raising capital finance for investments in water programmes, urban streets and electricity distribution, as well as replacing some high cost debt. Similarly, Pune in India has successfully managed to raise finance for urban water service delivery (see Box 6.1).

BOX 6.1 Pune's municipal bond issue for water services delivery

A 1994 amendment to the Indian constitution enabled urban authorities to generate revenue for infrastructure development and management. This has focused particularly on drainage, sewers and water, roads and transport. Under these arrangements, the Pune Municipal Corporation released a bond in 2017 to finance a water services project through investment in a reservoir, a distribution network and installation of water meters. The yield is higher than nationally issued bonds and is grounded on the city's financial strength, demonstrated ability to collect other taxation and good credit rating.

Sources: Govindarajulu (2020); Das (2017)

As such, there is growing potential for municipal bonds to be issued as green bonds; that is, as well as delivering a return on investment like a normal bond, some of the proceeds are explicitly earmarked for climate or environmental projects. The issuer then tracks and reports on use of the proceeds to ensure green compliance (CBI 2019). Green bonds thereby offer investors the same financial terms as other bonds, with the added bonus of delivering environmentally beneficial investments. The green bond market has been growing rapidly, with governments, multilateral banks and corporate entities issuing a total of USD 754 billion globally during 2007–2019 and a record USD 259 billion in 2019 across 49 different countries (CBI 2020). This rapid market growth reflects investors' growing awareness of climate issues and the benefits and impacts of green investments (Pham 2016).

Crucially, recent trends show the expansion of markets in Latin American and Asia-Pacific, after initial domination by European and North American markets. The Indonesian government issued the first Green Islamic Bonds (or Green Sukuk) in March 2018, raising USD 1.25 billion from conventional, Islamic and green investors. These are shariah-compliant bonds, where 100% of the proceeds finance green projects that contribute to goals such as tackling climate change and biodiversity preservation (UNDP 2018). Green bonds proceeds have provided finance for climate change and sustainability actions across a range of sectors (see Figure 6.3), with water management

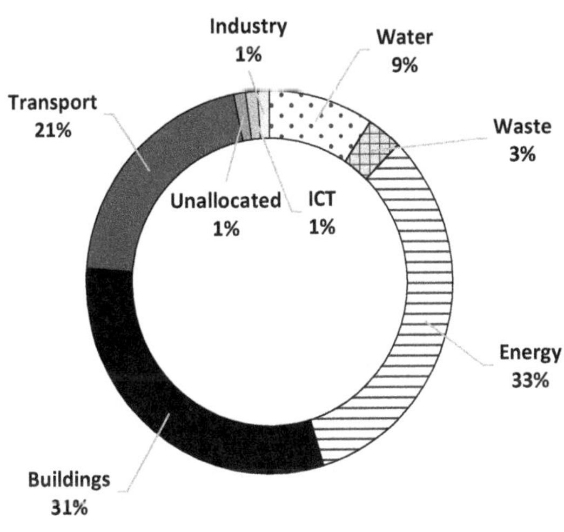

FIGURE 6.3 Use of proceeds by sector for green bond issues in 2019. Source: CBI (2020).

TABLE 6.2 Types of climate resilience investments for bond proceeds

Type of investment	Definition	Examples
Asset focused	The intention is to enhance the resilience of an asset or activity to climate change to ensure that the asset or activity is fit for purpose over its design lifespan	• Upgrading, replacing, relocating or adding spare capacity to critical and non-critical infrastructure • Using drought-resistant crops for agricultural production or training of farmers on sustainable farming practices
System focused	The investment is explicitly intended to deliver climate resilience benefits to the broader system	• Construction and operation of desalination plants • Wild brush clearing • Climate monitoring and data management technologies and services

Source: CBI (2019)

services and transport projects dominating the projects financed by municipalities to date (CBI 2020).

While climate change resilience may have been part of project goals incidentally, more recently there has been greater attention to climate change adaptation and resilience as specific criteria for the proceeds of green bonds. Climate change adaptation was one of the Green Bond Principles launched by a large group of international organisations in 2014, before the Climate Bonds Initiative led in drafting a framework for climate resilience principles in 2019 (CBI 2019; see Table 6.2). These distinguish between asset and system-focused investments, echoing the pivot in the previous chapter from investments in hard infrastructure to embracing softer areas such as data, technologies, tools, services or supply chains, that have a key role to play in enabling climate resilience.

Cape Town issued the first explicitly resilience-focused municipal green bond in 2017, and in 2018 the Environmental Finance Green Bond Awards named this Green Bond of the Year – local authority. Bond proceeds were allocated to improvements in reservoir and water treatment/distribution systems and flood protection measures in the city. Following such successes, Ngwenya and Simatele (2020) highlight the potential of green bonds to meet needs for climate financing across wider parts of the continent beyond the main economics hubs. To do so, they recommend changes to legal and regulatory frameworks to improve transparency and accountability from issuers; continuous monitoring and evaluation of the green bond market on the continent to determine investor behaviour and impacts; and encouraging political will and implementing policies to guide the green bond market.

The concentration of the green bond market in leading cities illustrates the multiple challenges to their wider diffusion to help finance urban resilience. First, adaptation and resilience are new fields for many urban actors, and meeting climate resilience and adaptation principles is relatively more complicated than planning for and reporting on, for example, GHG emissions reductions. Second, the ability of many city authorities in the Global South to issue bonds is fundamentally constrained by their existing low levels of creditworthiness. Third, there may be capacity and institutional constraints that influence the legal authority of municipalities over bonds. Drawing on evidence from sub-Saharan African cities, Gorelick (2018) concludes that the national legislative and political contexts of decentralisation are a major constraint on municipal bond success to date. He cites the example of Senegal's capital Dakar, where despite well-prepared plans at the city level, central government blocked a bond issue at the last minute, questioning both constitutional validity and the potential impact on national indebtedness. Gorelick also illustrates that successful bond issues are linked to the transparency, accountability and efficiency of more institutionalised democratic systems; non-transparent systems find it harder to tap into capital markets. Finally, green bond issues alone provide little guarantee that resulting financing will serve to enhance the resilience of the poorest and most vulnerable citizens For this it is crucial to have investment guidelines and frameworks that emphasise issues of equity and justice.

6.3.4 Innovative financing modalities: Climate risk insurance

There has been growing interest in the use of insurance as a means of managing climate change risks, and this provides an important entry point for cities in the future. Emerging approaches aim to demonstrate the benefits of using insurance to both share and transfer risks across wider areas and populations. This suggests that as well as compensating losses and funding recovery efforts, insurance instruments can also reduce the financial risk of investments and incentivise risk management activities (Schwarze et al. 2018). It also serves as a potential mechanism to tackle compound risks, as illustrated by the dual threats of the COVID-19 pandemic and climate-related disasters. Insurance can help enhance absorptive, adaptive and anticipatory capacities in a variety of ways (Schaefer and Waters 2016).

Payments against insurance premiums are mainly associated with helping individuals, businesses and governments absorb the impacts of shocks. Estimates suggest that indirect flood losses following the Mumbai floods of 2004 could have been halved at 100% insurance penetration rates (Ranger et al. 2011). This contrasts with the otherwise potentially destructive coping behaviours evident in times of crisis, both by individuals (such as selling assets or taking children out of educational settings) and organisations (such as municipal governments increasing their levels of indebtedness).

Insurance also helps to shift the emphasis of disaster financing towards anticipation, supported by improvements in data systems that track the severity of hazards and inform early warning systems to allow people to lessen impacts. Index-based systems that do not require verification of individual losses can provide support more rapidly than traditional post-disaster finance such as humanitarian aid or loans. For example, when drought hit Niger, Mauritania and Senegal in 2015, the Africa Risk Capacity insurance scheme paid out more than USD 26 million while a United Nations aid appeal was still in the process of being formulated (Schaefer and Waters 2016).

Finally, insurance has the potential to enhance adaptive capacities by fostering environments so that they are more conducive to making investment choices for climate resilience. At the micro level, insurance can boost productivity in years when there are no payouts (Greatrex et al. 2015). Similarly, providing coastal regions, towns, business districts or ports with protection from flood losses can foster economic activity, long-term planning and capital investments (Tanner et al. 2015). Crucially, holding insurance can enable firms and their stakeholders to make long-term capital investments and engage in trade, thus promoting business development and benefiting the entire urban area collectively. For example, Aerts and Botzen (2011) suggest that flood insurance schemes in New York and Rotterdam have reduced barriers to private investments in the waterfront and port areas. Demonstrating such proactive risk management can also potentially improve the creditworthiness of municipal authorities, urban businesses and households.

However, incentive structures need to be carefully managed to ensure that insurance promotes adaptive behaviours. First, insurance is prone to moral hazard, where policyholders make poor investment decisions in the knowledge that insurance will cover losses. Second, insurance needs to be part of a wider adaptation strategy that reduces risks more widely. If the root causes of rising risks, such as poor land use planning, are not addressed, insurance systems run the risk of being unsustainable due to either lack of supply (availability of cover) or demand (affordability of premiums) (Surminski et al. 2016). By the same logic, however, insurance can provide a price signal for the need to address underlying risks (Botzen et al. 2009).

One model, confusingly referred to as 'resilience bonds', has emerged as an option for financing urban resilience. These bonds build on the model of catastrophe bonds; rather than being structured as debt finance as per regular municipal bonds, catastrophe bonds act as an insurance policy, paying out to investors in the event of a disaster event above a certain predefined threshold. Resilience bonds model the difference in expected losses from a disaster event with or without a resilience-building project and monetises this value as a source of funding for those projects. Vaijhala and Rhodes (2018) liken resilience bonds to progressive health insurance programmes, but instead of incentivising healthy choices to reduce long-term care costs, they provide incentives to reduce disaster risks and therefore avoid future disaster losses. To date, resilience bonds have been confined mainly to North America, but they are of potential interest to municipal authorities, especially in contexts where there are larger asset holders with high insurance compliance requirements, communities seeking to expand the availability or improve the affordability of private property insurance, and cities with major resilience projects that lack funding (Re:bound 2018).

One major recent development at international level within insurance industries is the commitment to tackle barriers to providing insurance in poorer and more vulnerable contexts. The InsuResilience initiative announced by the G7 (2015) was launched to coincide with the Paris climate agreement in 2015 to bring climate insurance to 400 million new individuals in poor countries by 2020. Yet aside from economic considerations around affordability of premiums and ability to spread wide-reaching covariate risks, there are also ethical questions around shifting responsibility on to those who are the least responsible for climate change, the least able to pay insurance premiums and, in many cases, the least able to reduce their own losses (Surminski et al. 2016).

In tackling these challenges, bottom-up insurance schemes and the experiences of microcredit schemes may have potential where traditional insurance mechanisms are unable to deliver due to high premiums and entry barriers such as formal identification or proof of income. While there is no accepted definition of microinsurance, this commonly targets low-income populations and is characterised by low premiums. Yore and Walker (2019) review the application of 40 worldwide microinsurance initiatives to disaster management in the last 20 years, both successful and unsuccessful. Results suggest the importance of links to recognised national and international bodies, subsidising premiums to engage low-income clientele, portfolio risk sharing, design and implementation of initiatives, delivery channels to penetrate target markets and constructing stakeholder networks. Trust and credibility are vital, as illustrated by a study in Nairobi, Kenya, that found trust and capacity building play a dominant role in providing an enabling environment and creating a demand for urban index-based microinsurance to tackle flood risk in the city (Tarbuck 2018).

Such microinsurance schemes have a long track record of iterative testing and refinement in India. After the major Gujarat earthquake of 2001, a microinsurance scheme covering loss of life, trading stock, livelihood assets, home and home contents was first sold in 2004 to 3,700 policyholders; it was extended later to families affected by the 2004 Indian Ocean tsunami in Tamil Nadu, the 2005 earthquake in Jammu and Kashmir and later floods and a cyclone in Odisha. Crucially, policyholders were also supported with measures to reduce risk (such as fire safety actions, seismic-safe construction practices and business development) as well as awareness raising and education on disaster risk reduction through training, focus group discussions, dissemination of case studies and the creation of a platform to share ideas within the community. More recently, this approach has been extended to assess the demand for disaster microinsurance among informal small businesses (Patel et al. 2017).

In summary, a review for the InsuResilience programme (Schaefer and Waters 2016) suggests that successful pro-poor climate risk insurance schemes will rely on:

1. comprehensive needs-based solutions tailored to local needs and conditions, embedded in comprehensive risk management strategies that improve resilience;

2. provision of reliable coverage that is valuable to the insured;
3. affordability for poor and vulnerable, including to satisfy equity concerns;
4. efficient and cost-effective delivery channels that are aligned with the local context;
5. inclusive, meaningful and accountable involvement of (potential) beneficiaries and other relevant local-level stakeholders;
6. safeguarding of economic, social and ecological sustainability;
7. an enabling environment that accommodates and fosters pro-poor insurance solutions.

6.3.5 Innovative financing modalities: Land value capture

There is potential to widen revenue generation options through the further development of land use resilience charges and land value capture. Land value capture is a public financing mechanism whereby governments generate revenue through an increase in land or property value because of a regulatory decision (such as a change in development rights) or an investment in infrastructure or an amenity (World Bank 2018). They are then able to capture a dividend either through sale of property or land or by collecting taxes or charges, that offset the cost associated with those regulatory changes or investments.

Value can be captured variously by extending basic services to unused land, attracting developers and then charging for development rights, using land as collateral for the city to take on more debt, or charging taxes based on the services provided in particular areas. The principle of capturing added value by establishing greater mobility is well established in transport investment. Hong Kong's "rail plus property" model captures the uplift in property values in locations along new transit routes, Saõ Paulo has used similar instruments to raise over USD 1.2 billion in six years, and Curitiba is generating funds for mass transit corridors along with higher-density, mixed-use spaces and green areas (NCE 2014).

As the impact of climate shocks and stresses increasingly play out in urban environments, real estate markets within these areas will be affected. Levels of risk will likely become a more prominent element of competition within and between cities: "Neighbourhoods and whole cities that can demonstrate a lower susceptibility to the worst effects of climate change will begin to see their sustainability translated into market value" (Dunning and Lord 2020: 2). There is hence growing interest in whether this mechanism could generate revenues to finance improvements in urban resilience by capturing resilience dividends generated by reducing climate-related risks in those areas (Surminski and Tanner 2016).

Potential examples include improvements in public green space such as green corridors that enhance drainage and air quality at the same time as providing recreational and aesthetic benefits. These in turn serve to increase land and property values in those areas. Dividends from higher resulting land sales or property tax revenues can then be used by municipal authorities to finance resilience initiatives and incentivise activity in lower-risk areas. Higher land values or charges to individuals and businesses would be offset by higher property values, reduced flood risk to properties and lower resulting insurance premiums.

Schwarze et al. (2018) highlight the potential of investments in public space to help reduce risk as a key opportunity for land value capture. This could support financing for green infrastructure and nature-based solutions for resilience instead of the more expensive conventional infrastructure (Chausson et al. 2020). Examples include creating green roofs, establishing urban gardens, redesigning networked parks and introducing green corridors to improve runoff and retention of water to reduce flood risk. Co-benefits of such interventions include aesthetic qualities, clean air, reduced GHG emissions and reduced urban heat island effects. One assessment of green corridors in Cali, Colombia, suggests that such benefits can be significant (see Box 6.2).

A key challenge with this approach lies in accessing upfront finance, as costs are recovered in the medium or long term (Schwarze et al. 2018). In Saõ Paulo this challenge was overcome by issuing bonds to developers permitting relaxed planning regulations in a neighbourhood targeted for upgrading. The resulting USD 2.2 billion over ten years could then be used for essential infrastructure and housing in those areas, with the potential to finance resilient investment (Blanco et al. 2017; World Bank 2018). Another challenge lies in integrating sustainability and resilience concerns within land value capture mechanisms, not only as these considerations may be novel to policymakers, but also because these values speak to more dynamic and multiscale considerations than conventional approaches (Dunning and Lord 2020).

BOX 6.2 Land value capture for resilience investments in Cali, Colombia

Grafakos et al. (2019) explore land value instruments in relation to the benefits accrued by green infrastructure interventions for flood risk reduction through the implementation of corredores ambientales urbanos (environmental urban corridors). They analyse a range of resulting benefits, anticipating that land values will rise and that the increment can be captured to finance further investment on urban resilience.

The authors suggest that project implementation would increase land value from 0.6% to as high as 6.2% in some neighbourhoods, through planting of trees, open green spaces and vegetation, and development of pedestrian areas. In contrast to the negative experience of tax-based approaches elsewhere, they found that public-financed investment in effective green resilient infrastructure could be a catalyst for private investment in the area and development of related projects, which would, in theory, increase land values.

6.3.6 *Improving business cases: The triple dividend of resilience*

In addressing the challenge of bankability of resilience projects, there is considerable potential in the analytical approach of the triple dividend of resilience. The rationale for making investments to enhance resilience and adapt to climate change has traditionally been pitched around disaster risk reduction that minimises losses associated with climate-related shocks and stresses. These losses can pass indirectly to other areas. Among the most compelling arguments to invest in resilience is the ability to save lives. Early warning systems and cyclone shelters have been credited with huge reductions in death rates in Bangladesh: in 1970 Cyclone Bhola (Category 3) killed 300,000 people; in 1991 Cyclone Gorky (Category 4) killed 138,866 people; and in 1997 Cyclone Sidr (Category 5) killed 4,234 people (EM-DAT 2015). In the state of Odisha, India, 9,843 were killed in the supercyclone of 1999 (after receiving a two-day warning), compared to 47 people in 2013 when the similar in strength Cyclone Phailin hit the same area (after a four-day warning) (UNEP 2013; World Bank 2015).

Converting lives lost into economic values is problematic, but direct and indirect material losses can be presented to decision makers in terms of cost–benefit ratios or economic rates of return. For example, in the Mexican state of Tabasco between 2007 and 2010, the benefits of reduced flood damage due to investment in flood defence outweighed costs by a ratio of 4 to 1; in 2010 alone, flood losses of USD 3 billion, or 7% of the GDP of Tabasco, were avoided (World Bank, 2014). Even without economic valuation, reduction of physical loss can provide a powerful visible justification for investing in resilience. For example, homes built under the Storm-Resistant Housing for a Resilient Da Nang City project in Vietnam reported no damage when Typhoon Nari hit in October 2013 (Tran 2013).

Such approaches rely on demonstrating the costs of inaction. However, increasingly, the importance of financing urban resilience efforts is being framed not only in terms of protecting economic and social processes, assets, and citizen well-being, but also as a development opportunity. As illustrated in the examples of land value capture, there is growing evidence of the multiple co-benefits of actions to manage climate risks. The triple dividend of the resilience approach groups these according to three categories that improve the business case for investing in actions to enhance resilience (Tanner et al. 2015; see Figure 6.4).

First, avoidance of losses remains of paramount importance. It is common for this to be taken into account in assessments

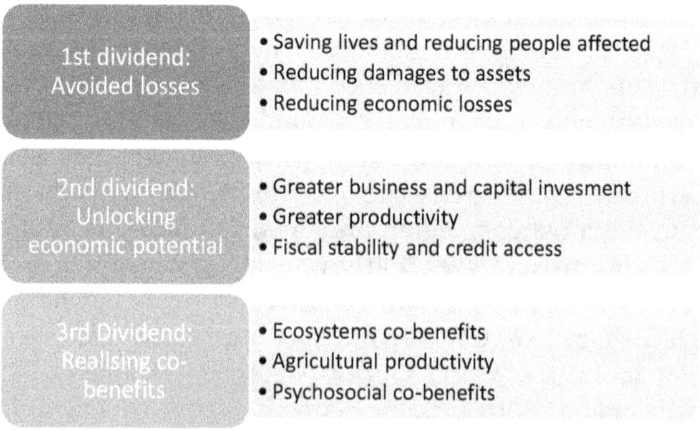

FIGURE 6.4 The triple dividend of resilience. Source: Adapted from Tanner et al. (2015).

of the cost–benefit ratio or return on investment for resilience building measures, but at the same time this can be difficult to calculate. Therefore, bankability needs to also incorporate the value of reduced "background risk" resulting from resilience initiatives, which can have immediate and significant development benefits. Reduced background risk enables longer-term capital investments and forward-looking planning, benefits which accrue even if disasters do not occur for many years. Benefits also include economic gains from positive risk-taking in terms of entrepreneurship and innovation, and greater investment in productive assets due to enhanced security against future shocks. This can increase land values, as illustrated in land value capture mechanisms (discussed in Section 6.3.5).

The benefits of lower background risk can also be reflected in better fiscal health and access to affordable credit for both governments and businesses (Tanner et al. 2015). There is evidence that climate change is already affecting sovereign creditworthiness through economic, fiscal and external performance (Buhr et al. 2018). Credit rating agencies are actively reporting that the credit profiles of both businesses and governments are likely to be tied more heavily to climate-related stresses and disaster events, and the exposure of global supply chains to such risks (Moody's 2015; Standard & Poor's Rating Services 2015). In the case of coastal cities, in the Hampton Roads region in south-eastern Virginia, United States, credit rating agencies have explicitly requested disaster loss prevention strategies in order for credit ratings to be maintained (Moody's 2015). Financing for urban resilience may therefore become explicitly encouraged by rating services, particularly to enable capital investment. As ratings agencies increasingly call for disclosure of firms' exposure to climate and disaster risks, this will incentivise both municipal governments and businesses to invest in building resilience (Standard & Poor's Rating Services 2015).

The third dividend emerges from the fact that investments have multiple uses; for instance, a flood retention lake that serves also as a recreational facility. Co-benefits may arise by chance, or they may be designed into the project. Such benefits are not due to reduced background risk; rather, they can occur even in the absence of disaster events. Such co-benefits can be economic (e.g. multi-use infrastructure); social (e.g. improved community cohesion); and environmental (e.g. upstream watershed protection). Where the design of projects intentionally integrates such co-benefits, this provides a stronger financial incentive to invest in urban resilience (Surminski and Tanner 2016; GCA 2019).

Innovation in urban resilience also has potential to generate economic co-benefits by transferring hardware and software as well as technical and organisational capacity to cities facing similar challenges (Schwarze et al. 2018).

By way of an example, flood management in Sri Lanka's capital Colombo demonstrates the range of benefits accruing from wetland protection and restoration. These wetlands help to reduce flood risk in the Colombo basin, but also secure livelihoods for residents through fishing and rice cultivation and provide opportunities for tourism and recreation. These areas also deliver a localised cooling effect of up to 10°C, compared to impervious streets. Nearby residents therefore save money on air-conditioning for buildings and homes. Other wetland co-benefits include waste water treatment, maintenance of freshwater supplies, carbon sequestration, climate regulation, water regulation, soil erosion regulation, pollination and nutrient cycling. Total benefits of wetland protection have been estimated at UDS 113 million to USD 127 million annually (Tanner et al. 2015).

6.4 CONCLUSIONS

This chapter suggests that financing for urban resilience needs to pivot away from a reliance on externally controlled sources and towards endogenous control. Underpinning this pivot is the need to ensure that needs of all citizens are considered and given a voice in decision-making. All the financing mechanisms and policies identified in this chapter have potentially inequitable outcomes. For example, the spending of municipal bonds may support infrastructure development that displaces rather than supports poor and marginalised people.

Even if innovative financing mechanisms can help to deliver improvements in municipal resilience, they will need to be accompanied by measures that more directly promote equity in the resulting resilience building initiatives. As such, many cities need to approach resilience financing with the needs of informal settlements and the informal sector in mind; they cannot rely on the formal private sector and centralised government initiatives alone to deliver equitable results (Satterthwaite et al. 2020). The reality for many urban residents is that municipal investments in resilience are either absent or do not benefit them, leaving significant residual risks borne by those who are often the most vulnerable citizens. In this regard, there is significant potential to draw on processes of autonomous innovation to support adaptation and resilience: bottom-up, locally situated, intuitive and iterative processes that contrast with more structured, expert-led and resource-intensive standardised business procedures (Bahadur and Doczi 2016). Understanding and implementing such processes can help draw in finance that more directly and effectively meets the needs of those most vulnerable to climate change impacts.

Addressing inequities and injustice therefore requires renewed attention to grassroots organisations and urban households in the city in the context of urban resilience financing. It also requires checks and balances to be built into resilience initiatives to assess and address trade-offs. Procedural considerations to promote voices across the range of urban stakeholders will also be vital. The shared learning dialogue approach used in the Asian Cities Climate Change Resilence Network programme illustrates how joint deliberation can promote debate and surface issues over the longer term (Archer and Dodman 2015).

Developing a justice- and rights-based framework for vulnerability and resilience in cities will therefore be a major critical endeavour in the future. This can be based on plausible (even if contested) views and needs to identify the underlying causes of why ideal justice and rights are not reflected in people's actual lived entitlements (Ziervogel et al. 2017). Normative embedding of these factors into emerging finance mechanisms, including via international frameworks and networks, can underpin the success of the pivot put forward here.

REFERENCES

Acclimatise. (2017). *EU law to force pension funds to account for climate risk.* Retrieved 23 November, 2020, from https://www.acclimatise.uk.com/2017/02/01/eu-law-to-force-pension-funds-to-account-for-climate-risk/

Aerts, J. C. J. H. and Botzen, W. J. (2011). Flood-resilient waterfront development in New York City: A study of flood insurance, building codes, and flood zoning. Annals of the New York Academy of Sciences, 1227, 1–82.

Alam, M. and Rabbani, G. (2007). Vulnerabilities and responses to climate change for Dhaka. *Environment and Urbanization, 19*(1), 81–97.

Alexander, J., Nassiry, D., Barnard, S., Lindfield, M., Teipelke, R., and Wilder, M. (2019). *Financing the Sustainable Urban Future: Scoping a Green Cities Development Bank*. London: Overseas Development Institute.

Allen, A., Griffin, L. and Johnson, C. (Eds). (2017). *Environmental Justice and Urban Resilience in the Global South*. New York: Palgrave Macmillan.

Archer, D. and Dodman, D. (2015). Making capacity building critical: Power and justice in building urban climate resilience in Indonesia and Thailand. *Urban Climate*, *14*(1), 68–78.

Avory, B., Cameron, E., Rickson, C. and Fresia, P. (2015). Climate Resilience and the *15* Role of the Private Sector in Thailand. Hong Kong: BSR Report.

Ayers, J. (2009). International funding to support urban adaptation to climate change. *Environment and Urbanization*, *21*(1), 225–240.

Bahadur, A. and Doczi, J. (2016). *Unlocking Resilience through Autonomous Innovation*. London: Overseas Development Institute.

Bahadur, A. and Tanner, T. (2014a). Policy climates and climate policies: Analysing the politics of building urban climate change resilience. *Urban Climate*, *7*, 20–32.

Bahadur, A. and Tanner, T. (2014b). Transformational resilience thinking: Putting people, power and politics at the heart of urban climate resilience. Environment and Urbanization, 26(1), 200–214.

Bahadur, A., Tanner, T. and Pichon, F. (2016). *Enhancing Urban Climate Change Resilience: Seven Entry Points for Action*. ADB Working Paper No. 47. Manila: Asian Development Bank.

Barnard, S. (2015). *Climate Finance for Cities: How Can International Climate Funds Best Support Low-Carbon and Climate Resilient Urban Development*. London: Overseas Development Institute.

Beltran, P. T. (2012). *International Financing Options for City Climate Change Interventions: An Introductory Guide*. Manilla: Cities Development Initiative for Asia.

Blanco, A. G., Moreno, N., Vetter, D. and Vetter, M. (2017). *The Potential of Land Value Capture for Financing Urban Projects: Methodological Considerations and Case Studies*. Washington, DC: Inter-American Development Bank.

Botzen, W. J. W., Aerts, J. C. J. H. and van den Bergh, J. C. J. M. (2009). *Willingness of homeowners to mitigate climate risk through insurance. Ecological Economics*, *68*(8–9), 2265–2277.

Buhr, B., Donovan, C., Kling, G., Lo, Y., Murinde, V., Pullin, N. and Volz, U. (2018). *Climate Change and the Cost of Capital in Developing Countries: Assessing the Impact of Climate Risks on Sovereign Borrowing Costs*. Geneva and London: UN Environment

Programme, Imperial College London and SOAS University of London.

C40. (2018). *The Demand for Financing Climate Projects in Cities: An Analysis of Projects from the C40 Cities Finance Facility's Application Phase and from CDP Disclosure*. London: C40 Cities Finance Facility, Carbon Disclosure Project, and Global Covenant of Mayors for Climate & Energy.

Chausson, A., Turner, B., Seddon, D., Chabaneix, N., Girardin, C. A., Kapos, V., Key, I., Roe, D., Smith, A., Woroniecki, S. and Seddon, N. (2020). Mapping the effectiveness of nature-based solutions for climate change adaptation. *Global Change Biology*, *26*(11), 6134–6155.

Chen, C., Doherty, M., Coffee, J., Wong, T. and Hellmann, J. (2016). Measuring the adaptation gap: A framework for evaluating climate hazards and opportunities in urban areas. *Environmental Science & Policy*, *66*, 403–419.

Chu, E. (2016). The political economy of urban climate adaptation and development planning in Surat, India. *Environment and Planning C: Government and Policy*, *34*(2), 281–298.

Chu, E., Brown, A., Michael, K., Du, J., Lwasa, S. and Mahendra, A. (2019). *Unlocking the Potential for Transformative Climate Adaptation in Cities. Background Paper*. Washington, DC and Rotterdam: Global Commission on Adaptation.

Cities Climate Finance Leadership Alliance. (2015) *State of City Climate Finance 2015*. New York: Cities Climate Finance Leadership Alliance.

Climate Bonds Initiative. (2019). *Climate Resilience Principles*. London: Climate Bonds Initiative.

Climate Bonds Initiative. (2020). *Green Bonds: The State of the Market 2018*. London: Climate Bonds Initiative.

Cook, M. J. and Chu, E. K. (2018). Between policies, programs, and projects: How local actors steer domestic urban climate adaptation finance in India. In S. Hughes, E. Chu and S. Mason (Eds), *Climate Change in Cities: Innovations in Multi-Level Governance*. Cham, Switzerland: Springer.

Crawford, M. and Seidel, S. (2013). *Weathering the Storm: Building Business Resilience to Climate Change*. Arlington, VA: Center for Climate and Energy Solutions.

Das, S. (2017, 17 June). *Pune raises Rs 200 crore in first municipal bond issue in 14 years*. Economic Times. Retrieved 23 November, 2020, from https://economictimes.indiatimes.com/news/economy/policy/pune-raises-rs-200-crore-in-first-municipal-bond-issue-in-14-years/articleshow/59222069.cms

Dunning, R. J. and Lord, A. (2020). Preparing for the climate crisis: What role should land value capture play? *Land Use Policy*, *99*, Article 104867.

EM-DAT. (2015). *The OFDA/CRED International Disaster Database*. Brussels: Université Catholique de Louvain.

Future Cities Catapult. (2014). *Urban Innovation and Investment: The Role of International Financial Institutions and Development Banks*. London: Future Cities Catapult.

G7. (2015). *Leaders' Declaration G7 Summit, 7–8 June 2015*. Retrieved 23 November, 2020, from https://sustainabledevelopment.un.org/content/documents/7320LEADERS%20STATEMENT_FINAL_CLEAN.pdf

Global Commission on Adaptation. (2019). *Adapt Now: A Global Call for Leadership on Climate Resilience*. Rotterdam and Washington DC: Global Center on Adaptation/World Resources Institute.

Gorelick, J. (2018). Supporting the future of municipal bonds in sub-Saharan Africa: the centrality of enabling environments and regulatory frameworks. *Environment and Urbanization*, *30*(1), 103–122.

Govindarajulu, D. (2020). Strengthening institutional and financial mechanisms for building urban resilience in India. International Journal of Disaster Risk Reduction, 47, Article 101549.

Grafakos, S., Tsatsou, A., D'Acci, L., Kostaras, J., Lopez, A., Ramirez, N. and Summers, B. (2019). *Exploring the Use of Land Value Capture Instruments for Green Resilient Infrastructure Benefits: A Framework Applied in Cali, Colombia*. Cambridge, MA: Lincoln Institute of Land Policy.

Greatrex, H., Hansen, J., Garvin, S., Diro, R., Blakeley, S., Le Guen, M., Rao, K. and Osgood, D. (2015). *Scaling Up Index Insurance for Smallholder Farmers: Recent Evidence and Insights*. CCAFS Report No. 14. Copenhagen: CGIAR Research Program on Climate Change, Agriculture and Food Security.

ICLEI. (2019) *Resilient Cities, Thriving Cities: The Evolution of Urban Resilience*. Bonn: ICLEI.

International Finance Corporation. (2018). *Climate Investment Opportunities in Cities*. Washington, DC: International Finance Corporation.

Jha, A., Miner, T. and Stanton-Geddes, Z. (Eds). (2013). *Building Urban Resilience: Principles, Tools, and Practice*. Washington, DC: World Bank.

Kennan, J. M., Chu, E. and Peterson, J. (2019). From funding to financing: Perspectives shaping a research agenda for investment in urban climate adaptation. *International Journal of Urban Sustainable Development*, *11*(3), 297–308.

Khattri, A., Parameshwar, D. and Pellech, S. (2010). *Opportunities for Private Sector Engagement in Urban Climate Change Resilience Building*. Hyderabad: Intellectual Capital Advisory Services.

Martín, C. and McTarnaghan, S. (2018). *Institutionalizing Urban Resilience: A Midterm Monitoring and Evaluation Report of 100 Resilient Cities*. New York: Urban Institute.

Matin, N., Forrester, J. and Ensor, J. (2018). What is equitable resilience? *World Development*, *109*, 197–205.

Moody's. (2015). *Flood Risk in Coastal Virginia Supports Need for Proactive Planning, Capital Investments*. New York: Moody's.

New Climate Economy. (2014). *Better Growth, Better Climate. The New Climate Economy Report*. London: NCE.

Ngwenya, N. and Simatele, M. D. (2020). Unbundling of the green bond market in the economic hubs of Africa: Case study of Kenya, Nigeria and South Africa. *Development Southern Africa*, *37*(6), 888–903.

Nightingale, A. J., Eriksen, S., Taylor, M., Forsyth, T., Pelling, M., Newsham, A., Boyd, E., Brown, K., Harvey, B., Jones, L., Bezner Kerr, R., Metha, L., Naess, L. O., Scooones, I., Tanner, T. and Whitfield, S. (2020). Beyond technical fixes: Climate solutions and the great derangement. *Climate and Development*, *12*(4): 343–352.

Patel, R., Walker, G., Bhatt, M. and Pathak, V. (2017). The demand for disaster microinsurance for small businesses in urban slums: The results of surveys in three Indian cities. *PLoS Currents*, *9*. Doi: 10.1371/currents.dis.83315629ac7cae7e2c4f78c589a3ce1c

Pham, L. (2016). Is it risky to go green? A volatility analysis of the green bond market. *Journal of Sustainable Finance and Investment*, *6*(4), 263–291.

Plastrik, P., Coffee, J. and Cleveland, J. (2019). *Playbook 1.0: How Cities Are Paying for Climate Resilience*. Innovation Network for Communities. Retrieved 23 November, 2020, from http://lifeaftercarbon.net/wp-content/uploads/2020/09/Playbook-1.0-How-Cities-Are-Paying-for-Climate-Resilience-July-2019-2.pdf

Ranger, N., Hallegatte, S., Bhattacharya, S., Bachu, M., Priya, S., Dhore, K. and Herweijer, C. (2011). An assessment of the potential impact of climate change on flood risk in Mumbai. *Climatic Change*, *104*(1), 139–167.

Re:bound. (2018). *A guide for public-sector resilience bond sponsorship*. Re:focus Partners. Retrieved 23 November, 2020, from www.refocuspartners.com/wp-content/uploads/pdf/RE.bound-Program-Report-September–2017.pdf

Satterthwaite, D., Archer, D., Colenbrander, S., Dodman, D., Hardoy, J., Mitlin, D. and Patel, S. (2020). Building resilience to climate change in informal settlements. *One Earth*, *2*(2), 143–156.

Schaefer, L. and Waters, E. (2016). *Climate Risk Insurance for the Poor and Vulnerable: How to Effectively Implement the Pro-Poor Focus of Insuresilience: An Analysis of Good Practice, Literature and Expert Interviews*. Bonn: Munich Climate Insurance Initiative.

Schwarze, R., Meyer, P. B., Markandya, A., Kedia, S., Maleki, D., Román de Lara, M. V., Sudo, T. and Surminski, S. (2018). Economics, finance, and the private sector. In C. Rosenzweig, W. Solecki, P. Romero-Lankao, S. Mehrotra, S. Dhakal and S. Ali Ibrahim (Eds), *Climate Change and Cities: Second Assessment Report of the Urban Climate Change Research Network* (pp. 225–254). New York: Cambridge University Press.

Standard & Poor's Rating Services. (2015, 20 April). *Ratings Direct: Climate Change Will Likely Test the Resilience of Corporates' Creditworthiness to Natural Catastrophes*. New York: Standard & Poor's Rating Services.

Surminski, S., Bouwer, L. M. and Linnerooth-Bayer, J. (2016). How insurance can support climate resilience. Nature Climate Change, 6(4), 333–334.

Surminski, S. and Tanner, T. M. (Eds). (2016). *Realising the Triple Resilience Dividend: A New Business Case for Disaster Risk Management*. Dordrecht: Springer International Publishing.

Tanner, T. M., Surminski, S., Wilkinson, E., Reid, R., Rentschler, J. E. and Rajput, S. (2015). *The Triple Dividend of Resilience: Realising Development Goals through the Multiple Benefits of Disaster Risk Management*. Washington DC and London: Global Facility for Disaster Reduction and Recovery at the World Bank and Overseas Development Institute.

Tarbuck, P. (2018). *An Exploratory Analysis into Urban Index-Based Microinsurance for Flood Risk in Nairobi: A Community-Based Approach*. Urban Africa Risk Knowledge Working Paper. London: Kings College London.

Tran Van Giai Phong. (2013). *Lessons from Typhoon Nari: Storm Resistant Housing Shown to Be Effective*. Hanoi: Institute for Social and Environmental Transition-International.

United Nations Development Programme. (2018) *Indonesia's Green Bond & Green Sukuk Initiative*. Ministry of Finance, Republic of Indonesia.

United Nations Environment Programme. (2013). *Cyclone Phailin in India: Early Warning and Timely Actions Save Lives*. Sioux Falls: UNEP, Global Environmental Alert Service.

United Nations Environment Programme. (2018). *The Adaptation Gap Report 2018*. Nairobi: United Nations Environment Programme.

Vaijhala, S. and Rhodes, J. (2018). Resilience bonds: A business-model for resilient infrastructure. *Field Actions Science Reports. The Journal of Field Actions*, Special Issue 18, 58–63.

Webber, S., Leitner, H. and Sheppard, E. (2020). Wheeling out urban resilience: Philanthrocapitalism, marketization, and local practice. *Annals of the American Association of Geographers*, *111*(2), 343–363.

Weru, J., Okoyo, O., Wambui, M., Njoroge, P., Mwelu, J., Otibine, E., Chepchumba, A., Wanjiku, R., Wakesho, T. and Njenga Maina, J. P. (2018). The Akiba Mashinani Trust, Kenya: A local fund's role in urban development. *Environment and Urbanization*, *30*(1), 53–66.

Wilkinson, E. (2012). Transforming *D*isaster *R*isk *M*anagement: A *P*olitical *E*conomy *A*pproach. Background Note. London: Overseas Development Institute.

World Bank. (2014). *A Novel Approach to Disaster Risk Management: The Story of Mexico*. Washington DC: World Bank.

World Bank. (2015). *Investing in Urban Resilience: Protecting and Promoting Development in a Changing World*. Washington DC: World Bank.

World Bank. (2018). *Financing a Resilient Urban Future: A Policy Brief on World Bank and Global Experience on Financing Climate-Resilient Urban Infrastructure*. Washington DC: World Bank.

Yore, R. and Walker, J. F. (2019). Microinsurance for disaster recovery: Business venture or humanitarian intervention? An analysis of potential success and failure factors of microinsurance case studies. *International Journal of Disaster Risk Reduction*, *33*, 16–32.

Ziervogel, G., Pelling, M., Cartwright, A., Chu, E., Deshpande, T., Harris, L., Hyams, K., Kaunda, J., Klaus, B., Michael, K., Pasquini, L., Pharoah, R., Rodina, L., Scott, D. and Zweig, P. (2017). Inserting rights and justice into urban resilience: A focus on everyday risk. *Environment and Urbanization*, *29*(1), 123–138.

Chapter 7

The urban resilience reset in a post-COVID-19 world

THE agenda presented in this book calls for a reset in thinking and practice, making the case for a series of pivots across the realms of data, communities, systems, capacities and finance. The reset metaphor is considered appropriate in part because urban resilience has reached a point of maturity; there exists significant research and practice globally on which to base the analysis presented in this book. We also make this call partly because the dangers of inertia and path dependency mean there is an urgent need to stop and take stock.

We take an explicitly normative position in prescribing a set of pivots based on our analysis across five domains. These pivots refer to the application of changes in the use of technologies, management approaches, different modes of engagement, but also changes to the approach to resilience in theory and practice. Conceptualisation is important as it influences the ways that resilience is framed as a problem and thereby limits the framing of the solutions that are proposed. Throughout, a normative underpinning to resilience emphasises rights, equity and justice considerations (Ziervogel et al. 2017; Matin et al. 2018).

The first chapter sets the context with trends in both urban development and climate risks. Rapid demographic and physical growth dominate urban trends in the Global South. Additional vulnerability in urban areas amplifies the impacts of climate hazards for several reasons, including the high density of human populations; dynamic migrant populations; high levels of poverty and inequality that limit adaptive capacity; large sections of the urban population living in informal housing, unregulated by land use controls and building standards; concentrations of solid and liquid wastes; impermeable surfaces and concentrations of buildings which disrupt natural drainage channels and enhance heat effects; and urban expansion onto particularly risky sites (Tanner et al. 2009). We then chart the evolution of urban resilience, including its diverse conceptual roots and the rapid growth and promotion of the term from the late 2000s onwards.

7.1 FIVE PIVOTS FOR URBAN RESILIENCE THINKING AND PRACTICE

The core chapters in the book tackle the five domains for the reset, while here we reflect on future directions and the evolving global resilience context in light of the COVID-19 pandemic. Chapter 2 critiques dominant modes of collecting, collating and deploying information on climate change risk and resilience. The predominance of climate models, satellite remote sensing, surveys and participatory approaches may be neither reliable nor appropriate in the context of the highly dynamic and granular realities of risk and resilience.

The first pivot is therefore a data pivot that explores how a new generation of approaches that draw on big data, machine learning, AI and innovative information and communication technologies can help bridge these gaps. Employing such novel, cost-effective and representative approaches alongside the existing suite of methods can improve our understanding and analysis of climate risk and resilience.

The second pivot, outlined in Chapter 3, critiques community-based approaches to climate change resilience, arguing that these approaches have commonly failed to engage with the underlying drivers of risk to deliver sustainable change at scale. We make a case for a pivot away from such incremental approaches to embrace transformational change that attempts to tackle the root causes of risk, deliver lasting change and catalyse broader changes while working at and across different scales of government.

A third pivot critiques the emphasis on formal urban planning mechanisms that have dominated many policy prescriptions for enhancing urban resilience. Master plans, city development plans and land use plans have helped build resilience by determining what, how and where development occurs in urban areas. We argue in Chapter 4 that the dominance of the informal economy and informal settlements in many cities render such planning approaches of limited use unless they embrace informal knowledge, actors and practices.

The systems approach that is emphasised in resilience thinking has provided an important pathway for building urban resilience. However, we argue in Chapter 5 that a disproportionate emphasis on the installation of techno-managerial solutions and hard infrastructure has come at the expense of investments in the capacity of institutions, organisations and individuals. A fourth pivot towards systems capacity is therefore needed to increase the authority, capability and technical capacity to manage risk and resilience appropriately in towns and cities.

In our fifth pivot, we address urban resilience financing (see Chapter 6). While externally generated finance has permitted an experimentation phase for urban resilience, we argue for approaches to unlock greater levels of endogenous finance that are controlled by governments, businesses and households within cities. We argue that this pivot requires greater attention to supportive enabling environments, upscaling of innovative financing tools, recognition of informality and equity, and focus on developing business cases that recognise and capture the multiple benefit streams of resilience building in urban areas.

The remainder of this final chapter pulls together the resilience reset in the context of complex and overlapping risks, exemplified by the COVID-19 pandemic. It then sets out the cross-cutting call to understand the politics underlying resilience building efforts in order to pursue normative agendas around justice and equity that form the backbone of the pivots outlined in the reset.

7.2 TACKLING COMPLEX URBAN RISKS: URBAN CLIMATE AND COVID-19 RESILIENCE

Urban areas are facing a complex mix of risks. While this book centres on climate-related risks in the context of a rapidly changing global climate, the ways that impacts are experienced, and how responses are managed, depend on the interaction of a complex wider range of risks. As noted in the opening chapter, part of the appeal of framing resilience in the urban context is the possibility that offers for understanding and managing risk more holistically. On a visit by the authors in 2018 to the Indian city of Surat, a municipal official illustrated this well; the official noted that while the floods of 2011 had heightened awareness of climate-related shocks, a core value in framing resilience involved this being set in the context of risk from shocks to the diamond trade, which dominates the city's economy.

Established patterns of risk are changing, statistically unlikely disturbances are occurring more frequently, and multiple crises are unfolding concurrently. In a globally connected world, risks can transfer across borders, while teleconnections transfer climate-related disturbances across the globe. The emergence of the COVID-19 pandemic and its interrelation with climate change risks provides a pertinent example of the complex crises now facing urban areas, with existing risk management systems and practices struggling to cope. We contend that while the COVID-19 crisis presents some specific challenges over and above climate change, the pivots identified across five domains in this book are instructive in tackling COVID-19 as part of complex risks in urban areas.

One important characteristic of the response to the COVID-19 crisis in many countries has been the turn to science and scientific information to inform policy responses. The recent climate change narrative of being "led by the science" has even been appropriated by governments in tackling COVID-19.

While this is not the case in all countries and all contexts, government respect for science is important because it should inform their decisions around testing, whether/how to reduce person-to-person transmission and the nature of health care provision as well as decisions that might influence eventual economic harm (Hamilton and Safford 2020). The extent to which citizens respect science is likely to influence how far they comply with any regulations that are established. Indeed, Brzezinski et al. (2020) demonstrate from mobile phone data records that the proportion of people who stay at home after lockdown policies go into effect is significantly lower in countries with higher concentrations of climate change sceptics.

The data pivot advocated in Chapter 2 can, therefore, be readily applied to the context of complex crises witnessed during the COVID pandemic, where large-scale and cost-effective data on highly dynamic infections and transmission are needed to inform policy decisions and regulations. Real-time big data from mobile phones has been central to "track and trace" systems that seek to isolate those with prolonged and/or close contact with infected individuals. China's Health Barcode smartphone app, that replaced paper-based access permits, was first developed in Hangzhou city before being rolled out to the rest of the country (Lin and Hou 2020). South Korea's much-lauded contact tracing system tracks infected individuals using a combination of security camera footage, credit card records, and GPS data from cars and mobile phones to trace the movement of individuals with COVID-19. One area of opportunity for synergies with urban climate change resilience is through standardisation of protocols for communication created through "smart cities" initiatives. This would enable better communication between cities and data platforms. Along with greater equity and transparency, this can enhance the potential for cooperation on disasters and resilience across multiple risks (Allam and Jones 2020).

Climate-related risks have the potential to greatly amplify COVID-19 transmission and morbidity through impacts such as direct damage to health facilities and housing, loss of livelihoods, health risks such as contaminated water, and greater congregation of people as they shelter in the aftermath of events (Okura et al. 2020). This is because "deep inequity, and associated exposure and vulnerability is another pivot that both drives and shapes governance responses to this [COVID-19] crisis" (Schipper et al. 2020: 4). The principles for transformational approaches to community resilience in urban areas have similarly read across to tackling health-related risks such as COVID-19.

In Chapter 3 we show that building resilience needs to tackle underlying root causes driving risk, rather than focusing only on proximate drivers of loss. For instance, there is a need to move beyond community-based flood early warning initiatives that exist to save lives and assets from flood events to address the underlying reasons why some groups of urban residents are forced to occupy flood-prone land in the first place. In her critique of the Habitat III Conference's New Urban Agenda, Maria Kaika (2017: 89) uses a medical analogy to suggest that path dependency on old indicators, smart cities and outmoded institutional frameworks "can only act as immunology: it vaccinates citizens and environments so that they can take larger doses of inequality and degradation in the future … but does little towards alleviating it".

As with urban climate resilience building actions, COVID-19 policies require similarly transformational objectives that recognise the structural drivers of risks and vulnerability. The environmental justice movement has long highlighted that African Americans in the United States are more likely to live in counties where federal air pollution standards are not met, leading to people's lungs being more vulnerable to the infection (Agyeman et al. 2003). This has combined with other factors that can concentrate exposure to the virus such as poor housing quality, indoor air pollution, poor ventilation and higher likelihood of jobs that can't be carried out from home. Comorbidity is therefore highest in the poorest and most polluted communities, evidenced by the higher levels of hospitalisation and deaths among African Americans and people of colour (Wilson et al. 2020). There is therefore an important opportunity in COVID-19 responses to tackle the root causes of risk, deliver lasting change and catalyse broader changes that can enhance resilience.

Reflecting the pivot set out in Chapter 4, there is already evidence from COVID-19 responses globally that embracing informality is necessary to effectively support urban centres. Studies in South Africa demonstrate the additional challenges faced in informal settlements in terms of lack of space to practice social distancing, overburdened infrastructure, lack of savings, loss of income and shortage of food, hunger and diseases, anxiety and depression and poor access to education (Nyashanu et al. 2020: 1443).

When cases of the virus started rising in Dharavi in Mumbai – Asia's largest informal settlement and among the most densely populated places in the world – it was expected to spread out of control in the face of limited handwashing, cramped living conditions and high underlying comorbidity. Yet the success of its "chase the virus" strategy has been lauded as an example for other informal urban areas.

The Dharavi success was achieved through a proactive and holistic response, coordinating local support teams and recognising informal processes in the area (Figure 7.1). The municipal corporation cordoned off the infection centres and disinfected the 425 public toilets, and local teams moved from house to house to explain the virus, look for patients with symptoms and offer quarantine facilities. Crucially, NGO partners were pivotal in building community trust and providing food and in recognition of the psychosocial impacts of the virus through loss of income, savings and loved ones, making individuals vulnerable to anxiety, depression, panic attacks and other hitherto latent symptoms (Parasuraman 2020). As such, the Dharavi model has been held up globally as a template for policymakers and public health practitioners due to its success in breaking the chain of transmission in densely packed urban

FIGURE 7.1 Tackling COVID-19 in Dharavi, Mumbai.
Source: Adapted from Golechha (2020)

slum communities where social distancing is almost impossible (Golechha 2020).

The COVID-19 pandemic has served to illustrate the importance of improving human capacity to manage risk and uncertainty in urban areas. As highlighted in Chapter 5, despite the improvements in urban resilience, there has been an over-reliance on techno-managerial approaches and hard infrastructure, reflecting wider biases in climate adaptation and resilience building efforts (Nightingale et al. 2020). While news stories have enjoyed reporting on the engineering feats of rapidly constructed hospitals and the development of smart apps, the types of capacity development called for in our Chapter 5 pivot are also vital. The appearance of a pandemic that few had predicted at such great scale and speed demonstrates the need to live with and manage uncertainty, both known and unknown. Given the uncertain nature of risk that cities face, enhancing the ability of those with a part in running cities to manage and make decisions under uncertainty is crucial. The dynamism of the changing risks facing the world clearly demonstrate how "consideration of future states can be useful to inform societal and governance choices at critical junctures that lead to alternative development pathways" (Schipper et al. 2020: 3).

Finance for building urban climate resilience is now similarly bound up in COVID-19 response and recovery. As the case made in Chapter 6 sets out, a pivot towards endogenously driven finance is necessary to tackle both crises. Slow-moving public finance from national and international sources is less useful for dealing exigent problems. Sources of international aid, or official development assistance, for resilience are also under threat from both the switch to COVID-19 responses and anticipated reductions in assistance, as pledges are commonly tied to levels of GDP, which are being suppressed by the pandemic. This makes the need to develop joint responses even more acute, and there are some grounds for optimism that COVID-19 recovery plans can be integrated into low-carbon and resilient development plans, building on existing efforts (Hepburn et al. 2020). Quevedo et al. (2020) highlight public services, infrastructure, nature-based solutions and ecotourism as areas of potential for financing synergies.

At the same time, existing urban resilience financing mechanisms can be widened to encompass the complex crises exemplified by the pandemic. This reflects the need for a financing pivot, as outlined in Chapter 6, and innovative endogenous mechanisms in response to the complex crisis of climate and pandemic risks. In the United States, for example, the Federal Reserve bought municipal bonds to aid COVID-19 recovery, permitting locally led financing of the response in urban areas (Cheng et al. 2020). The emergence of forecast-based financing mechanisms provides one avenue for such integration, while locally managed social protection mechanisms are already adapted to targeting socially and economically marginalised citizens (Lind et al. 2020).

7.3 THE POLITICS OF URBAN CLIMATE CHANGE RESILIENCE

As outlined in the initial chapter, underpinning the resilience reset across the pivots is a call to understand the politics underlying resilience building efforts. Resilience is an inherently political enterprise. This can be understood by examining four key questions common to a range of scholars: Resilience of what?

To what? For whom? By whom? (Carpenter et al. 2001; Davoudi et al. 2012; Cretney 2014; Meerow and Newell 2019).

7.3.1 Resilience of what?

Crucial to any process to enhance resilience is a discussion on *what* is being made resilient. Determining the object or subject of resilience building exercises entail choices that are political. For example, in the aftermath of Hurricane Katrina, New Orleans has attempted to enhance its resilience by rebuilding public infrastructure and housing so as to withstand a diversity of shocks and stresses. Despite this, some argue that this vision of resilience is spurious as

> judging resilience depends on where one looks, given that the city's repopulation has taken radically different forms from neighbourhood to neighbourhood in the years since the August 2005 disaster. Is "New Orleans" resilient even if some of its component neighbourhoods remain half-empty? Is "the city" resilient even if many of its poorest former citizens have not been able to return?
>
> (Vale 2014: 197)

Another dimension of this problem involves trade-offs in structure and function in what is made resilient. For instance, a city may choose to enhance the resilience of the power supply to financially important neighbourhoods, but it may do this by adding additional coal-fired power plants in peri-urban areas (inhabited by marginalised groups) that exacerbate pollution and, over the long term, contribute to global warming and local respiratory disease (Smith and Stirling 2010). In this way, questions of *what* is being made resilient and *how* it is being made resilient need to be deliberated by key constituencies with a stake in ensuring the health, well-being and resilience of the city (Bahadur and Thornton 2013).

7.3.2 Resilience to what?

An urban system is subject to different sets of shocks and stresses. As illustrated by the shifting of funds during the COVID-19 crisis, selecting which hazards to prioritise is a political choice or decision (Quevedo et al. 2020). This choice has equity and justice implications. For example, for communities living in informal settlements and working in street markets, windstorms may be an overwhelming issue, while wealthier residents may be more concerned with extreme rainfall events. Therefore, the hazards that are prioritised as part of any process to build urban resilience could lead to more benefits for certain sections of the population than others. Crucially, actions to enhance resilience to one particular hazard may result in raising the vulnerability or exposure of the system to another. For instance, the Government of Vanuatu has undertaken construction of "hybrid classrooms" built with special, lightweight materials that make schools more resilient to earthquakes, but this also makes the classrooms more susceptible to damage from cyclones (Peters and Bahadur 2016).

7.3.3 Resilience for whom?

Efforts to enhance resilience will deliver different outcomes for some groups over others. We cannot consider resilience without "paying attention to issues of justice and fairness in terms of both the procedures for decision-making and the distribution of burdens and benefits" (Davoudi et al. 2012: 306). Different groups will benefit from actions to ensure an urban transport system can withstand climate risks and actions aimed at ensuring markets are able to re-open swiftly after disturbances, ensuring resilient livelihoods for street vendors. Weighing the relative costs and benefits of these outcomes for different groups and managing trade-offs between such choices are central to equity and justice considerations. Conversely, failure to take these considerations into account may lead to resilience initiatives that reinforce existing urban inequalities. Processes of enhancing resilience may in fact entail a redistribution of risk and resilience for other groups of urban residents. For instance, in the Indian city of Gorakhpur, residents decided that building walls around their compounds would reduce damage from waterlogging, a form of low-level, high-frequency urban flooding. However, a group of residents objected to this action as it would mean that flood waters would flow past the walled compounds and into their houses, that were located in poorer neighbourhoods downstream. This led to a rethinking of measures to enhance resilience to ensure that benefits were realised more equitably (Bahadur and Tanner, 2014).

7.3.4 Resilience by whom?

Much discourse on resilience has questioned who controls those efforts. The emphasis in resilience on preparedness, adaptability and "bouncing back" have been critiqued as encouraging active citizenship, whereby people are asked to take responsibility for their own social and economic well-being, rather than the state taking responsibility for this (Joseph 2013). Rinne and Nygren (2016) similarly argue that framing the problem of urban flooding in Mexico in terms of resilience has facilitated the view that battling floods is less a responsibility of the state and more about "self-responsibility" and "self-governance". For some, the emphasis on individuals as resilient subjects "represents a new phase in the neoliberal shift from the state as provider to state as enabler and promoter of self-reliance" (Mckeown and Glenn 2018: 193).

Despite calls for bottom-up approaches, we acknowledge that while actions by households and communities can increase resilience, the effectiveness of these actions can depend on complementary action by government (Satterthwaite 2013). The question of resilience by whom also pertains to decision-making. Resilience relies on an understanding of how different elements of a system interact and relate to each other (Simonsen 2015). In the context of towns and cities, this understanding can only result from the active participation in decision-making of groups that understand how different parts of the urban system function. Excluding some perspectives from decision-making processes will at best result in an inadequate understanding of interactions and feedbacks between different system components essential for ensuring resilience and at worst deliver inequitable outcomes from processes of building resilience.

7.4 A FORWARD-LOOKING AGENDA

This book makes the case for a resilience reset that takes stock of conceptualisation, policy and action to date in tackling urban resilience. While the COVID-19 pandemic has taken centre stage in terms of global risks since 2020, the ongoing incidence of disaster events linked to natural hazards has been a firm reminder of the importance of climate resilience in urban areas. The devastation caused in 2020 by Supercyclone Amphan in Bangladesh and India, from Supertyphoon Goni in the Philippines, and Hurricane Eta across Central America all attest to the urgency of actions for climate resilience.

The pivots presented in this book set out a framework for reset and for future enquiry and action. Across these pivots, there is urgent need for enquiry in three spheres. First, while we make the case for pivots, there is a need for greater understanding of the conditions in which these shifts can thrive and evolve. This involves understanding the institutional, behavioural, social and economic factors that may facilitate or inhibit pivots. Understanding the political economy that frames the understanding of problems and the deployment of potential solutions is central to this enquiry (Tanner and Allouche 2011) and reflects how different narratives of urban resilience can employ knowledge and science in different ways to legitimate or alienate particular perspectives on what should be done (Borie et al. 2019).

Second, normative drivers of equity and justice must steer the resilience reset. This may be guided by a multilateral system such as the Sendai Framework for Disaster Risk Reduction or be locally driven, as was the case in Durban in South Africa (Roberts et al. 2020). The resilience concept has itself been critiqued for an absence of normative framing due to its multiple ontological foundations and consequent difficulties of translation from the physical to social sciences (Béné et al. 2012; Olsson et al. 2015). We share the assertion of Matin et al. (2018) that the construction of equitable resilience is more likely to emerge when practice takes on board issues of social vulnerability and differential access to power, knowledge and resources. Integrating subjective perceptions of resilience provides one way of building accounts that reflects such differences (Jones and Tanner 2017).

Finally, we were conscious in the research and writing of this book of the need for decolonising knowledge and practices of urban resilience. This is not to throw out the normative goals we set out herein; the need to rework resilience from the inside out is exemplified in our pivots on community-based transformations, local capacity development, embracing informality and endogenous financing. But fundamentally we recognise that many of the terms and concepts supporting resilience agendas have been framed and developed to a large extent by elite groups of researchers and practitioners who are largely external to the urban centres of the Global South (Webber et al. 2020). The next generation of "post-reset" urban resilience research and practice therefore needs to engage the bottom-up perspectives and knowledges that are free from global narratives and more attuned and appropriate to city needs and capacities. Resilience can thereby include "both systemic and more situated and endogenous notions of resilience, where systems create, or build on and enhance, people's own capacity and resilience" (Ziervogel et al. 2017: 133).

REFERENCES

Agyeman, J., Bullard, R. D. and Evans, B. (Eds). (2003). *Just Sustainabilities: Development in an Unequal World*. London: Earthscan.

Allam, Z. and Jones, D. S. (2020). On the coronavirus (COVID–19) outbreak and the smart city network: Universal data sharing standards coupled with artificial intelligence (AI) to benefit urban health monitoring and management. *Healthcare*, *8*(1), 46.

Bahadur, A. and Tanner, T. (2014). Transformational resilience thinking: Putting people, power and politics at the heart of urban climate resilience. *Environment and Urbanization*, *26*(1), 200–214.

Bahadur, A. and Thornton, P. (2013). *Reimagining Resilience: Bringing Resilience, Transformation and Vulnerability Closer for Tackling Climate Change*. Bangkok: Rockefeller Foundation.

Béné, C., Wood, R. G., Newsham, A. and Davies, M. (2012). Resilience: New utopia or new tyranny? Reflection about the potentials and limits of the concept of resilience in relation to vulnerability reduction programmes. *IDS Working Papers*, *2012*(405), 1–61.

Borie, M., Pelling, M., Ziervogel, G. and Hyams, K. (2019). Mapping narratives of urban resilience in the Global South. *Global Environmental Change*, *54*, 203–213.

Brzezinski, A., Kecht, V., Van Dijcke, D. and Wright, A. L. (2020). *Belief in Science Influences Physical Distancing in Response to Covid-19 Lockdown Policies*. Becker Friedman Institute for Economics Working Paper 2020-56. Chicago: University of Chicago.

Carpenter, S., Walker, B., Anderies, J. M. and Abel, N. (2001). From metaphor to measurement: Resilience of what to what? *Ecosystems*, *4*(8), 765–781.

Cheng, J., Skidmore, D. and Wessel, D. (2020). *What's the Fed doing in response to the COVID-19 crisis? What more could it do?* Washington DC: Brookings Institution. Retrieved 23 November, 2020, from https://www.brookings.edu/research/fed-response-to-covid19/

Cretney, R. (2014). Resilience for whom? Emerging critical geographies of socio-ecological resilience. *Geography Compass*, *8*(9), 627–640.

Davoudi, S., Shaw, K., Haider, L. J., Quinlan, A. E., Peterson, G. D., Wilkinson, C., Fünfgeld, H., McEvoy, D., Porter, L. and Davoudi, S. (2012). Resilience: A bridging concept or a dead end? *Planning Theory & Practice*, *13*(2), 299–333.

Golechha, M. (2020). COVID–19 containment in Asia's largest urban slum Dharavi-Mumbai, India: Lessons for policymakers globally. *Journal of Urban Health*, *97*(6), 796–801.

Hamilton, L. C. and Safford, T. G. (2020). *Ideology Affects Trust in Science Agencies During a Pandemic*. Carsey Perspectives. Durham: Carsey School of Public Policy at the University of New Hampshire.

Hepburn, C., O'Callaghan, B., Stern, N., Stiglitz, J. and Zenghelis, D. (2020). Will COVID–19 fiscal recovery packages accelerate or retard progress on climate change? *Oxford Review of Economic Policy*, *36*(S1), S359–S381.

Jones, L. and Tanner, T. (2017). "Subjective resilience": Using perceptions to quantify household resilience to climate extremes and disasters. *Regional Environmental Change*, *17*(1), 229–243.

Joseph, J. (2013). Resilience as embedded neoliberalism: A governmentality approach. *Resilience: International Policies, Practices and Discourses*, *1*(1), 38–52.

Kaika, M. (2017). "Don't call me resilient again!" The New Urban Agenda as Immunology … or … what happens when communities refuse to be vaccinated with "smart cities" and indicators. *Environment and Urbanization 29*(1): 89–102.

Lin, L. and Hou, Z. (2020). Combat COVID–19 with artificial intelligence and big data. *Journal of Travel Medicine*, *27*(5), Article taaa080.

Lind, J., Roelen, K. and Sabates-Wheeler, R. (2020). Social Protection and Building Back Better. Positioning Paper. Brighton: Institute of Development Studies.

Matin, N., Forrester, J. and Ensor, J. (2018). What is equitable resilience? *World Development*, *109*, 197–205.

Mckeown, A. and Glenn, J. (2018). The rise of resilience after the financial crises: A case of neoliberalism rebooted? *Review of International Studies*, *44*(2), 193–214.

Meerow, S. and Newell, J. P. (2019). Urban resilience for whom, what, when, where, and why? *Urban Geography*, *40*(3), 309–329.

Nightingale, A. J., Eriksen, S., Taylor, M., Forsyth, T., Pelling, M., Newsham, A., Boyd, E., Brown, K., Harvey, B., Jones, L., Bezner Kerr, R., Mehta, L., Naess, L. O., Ockwell, D., Scoones, I., Tanner, T. and Whitfield, S. (2020). Beyond technical fixes: Climate solutions and the great derangement. *Climate and Development*, *12*(4), 343–352.

Nyashanu, M., Simbanegavi, P. and Gibson, L. (2020). Exploring the impact of COVID–19 pandemic lockdown on informal settlements in Tshwane Gauteng Province, South Africa. *Global Public Health*, *15*(10), 1443–1453.

Okura, Y., Dutta, S., Begum, A. and Naznin, Z. (2020). *Monsoon, floods and COVID-19: Building community resilience in Bangladesh*. Zurich Flood Resilience Alliance.

Olsson, L., Jerneck, A., Thoren, H., Persson, J. and O'Byrne, D. (2015). Why resilience is unappealing to social science: Theoretical and empirical investigations of the scientific use of resilience. *Science Advances*, *1*(4), Article e1400217.

Parasuraman, S. (2020, 25 June). *COVID–19: Involving social workers key to building on early gains in Dharavi*. Times of India. Retrieved 23 November, 2020, from https://www.thehindu.com/opinion/op-ed/covid-19-involving-social-workers-key-to-building-on-early-gains-in-dharavi/article31911787.ece

Peters, K. and Bahadur, A. (2016). *The Case of Vanuatu. Advancing Integration of Disaster, Environment and Climate Change*. London: Overseas Development Institute.

Quevedo, A., Peters, K. and Cao, Y. (2020). *The impact of Covid-19 on climate change and disaster resilience funding*. Briefing Note. London: Overseas Development Institute.

Rinne, P. and Nygren, A. (2016). From resistance to resilience: Media discourses on urban flood governance in Mexico. *Journal of Environmental Policy & Planning*, *18*(1), 4–26.

Roberts, D., Douwes, J., Sutherland, C. and Sim, V. (2020). Durban's 100 resilient cities journey: Governing resilience from within. *Environment and Urbanization*, *32*(2), 547–568.

Satterthwaite, D. (2013). The political underpinnings of cities' accumulated resilience to climate change. Environment and Urbanization, 25(2), 381–391.

Schipper, L., Eriksen, S E., Fernandez Carril, L. R., Glavovic, B. C. and Shawoo, Z. (2020). Turbulent transformation: Abrupt societal disruption and climate resilient development. *Climate and Development* [Advance online publication]. https://doi.org/10.1080/17565529.2020.1799738

Simonsen, S. H. (2015). *Applying Resilience Thinking: Seven Principles for Building Resilience in Social-Ecological Systems*. Stockholm: Stockholm Resilience Centre.

Smith, A. and Stirling, A. (2010). The politics of social-ecological resilience and sustainable socio-technical transitions. *Ecology and Society*, *15*(1), Article 11. www.ecologyandsociety.org/vol15/iss1/art11/

Tanner, T. and Allouche, J. (2011). Towards a new political economy of climate change and development. *IDS Bulletin*, *42*(3), 1–14.

Tanner, T., Mitchell, T., Polack, E. and Guenther, B. (2009). *Urban Governance for Adaptation: Assessing Climate Change Resilience in Ten Asian Cities*. IDS Working Papers 315. Brighton: Institute of Development Studies.

Vale, L. J. (2014). The politics of resilient cities: Whose resilience and whose city? *Building Research & Information*, *42*(2), 191–201.

Webber, S., Leitner, H. and Sheppard, E. (2020). Wheeling out urban resilience: Philanthrocapitalism, marketization, and local practice. *Annals of the American Association of Geographers*, *111*(2), 343–363.

Wilson, S., Bullard, R., Patterson, J. & Thomas, S. B. (2020). Environmental Justice Roundtable on COVID–19. *Environmental Justice*, *13*(3): 56–64.

Ziervogel, G., Pelling, M., Cartwright, A., Chu, E., Deshpande, T., Harris, L., Hyams, K., Kaunda, J., Klaus, B., Michael, K., Pasquini, L., Pharoah, R., Rodina, L., Scott, D. and Zweig, P. (2017). Inserting rights and justice into urban resilience: A focus on everyday risk. *Environment and Urbanization*, *29*(1), 123–138.

Index

Note: Page numbers in *italics* indicate figures and in **bold** indicate tables on the corresponding pages.

www.ingramcontent.com/pod-product-compliance
Lightning Source LLC
LaVergne TN
LVHW011819200926
848484LV00005B/1241
9780367375485